Malaysia-Japan Workshop on Radio Technology (MJWRT2022)

Journal of Physics: Conference Series Volume 2250

Online
11 – 12 January 2022

ISBN: 978-1-7138-5504-0
ISSN: 1742-6588

Printed from e-media with permission by:

Curran Associates, Inc.
57 Morehouse Lane
Red Hook, NY 12571

Some format issues inherent in the e-media version may also appear in this print version.

Institute of Physics
Dirac House, Temple Back
Bristol BS1 6BE UK

Phone: 44 1 17 929 7481
Fax: 44 1 17 920 0979

techtracking@iop.org

Additional copies of this publication are available from:

Curran Associates, Inc.
57 Morehouse Lane
Red Hook, NY 12571 USA
Phone: 845-758-0400
Fax: 845-758-2633
Email: curran@proceedings.com
Web: www.proceedings.com

TABLE OF CONTENTS

MJWRT2022 Preface

The Malaysia-Japan Workshop on Radio Technology 2022 (MJWRT2022) has been successfully hosted virtually on 11th and 12th January 2022 from Kuala Lumpur, Malaysia. Due to Covid-19 pandemic restrictions, MJWRT2022 was held virtually.

The MJWRT2022 is sponsored and organized by the Technical Committee on Antennas and Propagation of the Institute of Electronics, Information and Communication Engineers (IEICE/AP), the IEEE Malaysia AP/MTT/EMC Joint Chapter and the IEEE AP-S Tokyo Chapter, and supported by the Malaysia-Japan International Institute of Technology, Universiti Teknologi Malaysia (MJIIT-UTM). Comrel Technologies Sdn Bhd, 7Gaa Ltd. And Altair Engineering were also contributed as industry sponsors.

The workshop has successfully gathered 50 attendees from Japan and Malaysia which include academia and industry participants. The Dean of MJIIT, Prof. Dr. Ts. Ali Selamat officiated the workshop during the opening ceremony.

Delegates from Malaysia were represented by Associate Prof. Dr Azremi Abdullah Al-Hadi, the Chair of IEEE Malaysia AP/MTT/EMC while delegates from Japan were led by Prof. Dr. Hiroyoshi Yamada, the Chair of IEICE/AP Japan (Technical Committee on Antennas and Propagation). Industry speakers from Altair Engineering also participated in the workshop.

MJWRT2022 is the fourth workshop on Radio Technology between Malaysia and Japan. The workshop is intended to provide an international forum for the exchange of information on the progress of research and development in antennas, propagation, Radio Frequency (RF) and radio communication systems.

An equally important objective of this workshop is to enhance the friendship between Malaysian and Japanese researchers.

ORGANIZING COMMITTEE

General Co-Chairs

- Prof. Dr. Hiroyoshi Yamada (Niigata University, Japan)
- Assoc. Prof. Dr. Azremi Abdullah Al-Hadi (Universiti Malaysia Perlis, Malaysia)

Vice Chairs

- Prof. Dr. Yoshihide Yamada (Malaysia-Japan International Institute of Technology, Universiti Teknologi Malaysia, Malaysia)
- Prof. Dr. Mohamad Kamal A. Rahim (Universiti Teknologi Malaysia, Malaysia)
- Prof. Mitoshi Fujimoto (Fukui University, Japan)

Financial Chairs/ IEEE Liaison

- Ir. Dr. Kamilia Kamardin (Malaysia-Japan International Institute of Technology, Universiti Teknologi Malaysia, Malaysia)

Technical Program Chair

- Dr. Hazilah Mad Kaidi (Universiti Teknologi Malaysia, Malaysia)
- Prof. Takashi Hikage (Hokkaido University, Japan)

Publicity Chair

- Ir. Dr. Nurul Huda Abd. Rahman (Uniersiti Teknologi MARA, Malaysia)

Technical & Website Chair

- Assoc. Prof. Ir. Dr. Nur Emileen Abd Rashid (Universiti Teknologi MARA, Malaysia)

Local Arrangement Chair

- Dr. Sumiaty Ambran (Malaysia-Japan International Institute of Technology, Universiti Teknologi Malaysia, Malaysia)

Secretary

- Dr. Azura Hamzah (Malaysia-Japan International Institute of Technology, Universiti Teknologi Malaysia, Malaysia)
- Prof. Kentaro Saito (Tokyo Denki University, Japan)

Peer Review Statement

All papers published in this volume have been reviewed through processes administered by the Editors. Reviews were conducted by expert referees to the professional and scientific standards expected of a proceedings journal published by IOP Publishing.

- **Type of peer review:** Single Anonymous
- **Conference submission management system:** Morressier
- **Number of submissions received: 17**
- **Number of submissions sent for review: 17**
- **Number of submissions accepted: 16**
- **Acceptance Rate (Submissions Accepted / Submissions Received × 100):** 94
- **Average number of reviews per paper:** 2
- **Total number of reviewers involved:** 28
- **Contact person for queries:**
  Name: HAZILAH MAD KAIDI
  Email: hazilah.kl@utm.my
  Affiliation: UNIVERSITI TEKNOLOGI MALAYSIA

Frequency Integration of Dual-band Hexagonal Metamaterial Resonator antenna for Wi-Fi and 5G Wireless Communication

M L Hakim [1], T Alam[1,2] , M H Baharuddin[3] and M T Islam[3]

[1]Space Science Centre (ANGKASA), Institute of Climate Change (IPI), Universiti Kebangsaan MalaysiaBangi 43600, Selangor, Malaysia.
[2]Dept. of CSE, International Islamic University Chittagong (IIUC) Chittagong, Bangladesh
[3]Dept. of Electrical, Electronic and Systems Engineering, Universiti Kebangsaan Malaysia Bangi 43600, Selangor, Malaysia
Email: touhidul@ukm.edu.my, tariqul@ukm.edu.my

Abstract. The integration of 5G mm-wave and Wi-Fi frequency bands becomes significant for upcoming 5G wireless applications for providing a large frequency ratio. This work presents a frequency reconfigurable hexagonal metamaterial resonator (HMR) antenna for Wi-Fi and a 5G mm-wave frequency band. A PIN diode is utilized to link up a microstrip patch antenna (MPA) with HMR to achieve integration of the Wi-Fi frequency band. The resonant frequency of the proposed antenna is 5 GHz (Wi-Fi) and 27.87 GHz with peak gain 2.33 dB and 7.12 dB, respectively. The operating bandwidth of the proposed antenna is 800 MHz (4.5-5.3 GHz) and 1020 MHz (27.35-28.37 GHz). The rogers RT5880 substrate with thickness 0.508 mm, dielectric constant 2.2, and loss tangent 0.009 has been used for designing the antenna. The overall size of the antenna is $30\times23\times0.508$ mm^3.

1. Introduction

5G or fifth-generation wireless communication with the feature of lower latency, high-level data rate, and outstanding reliability facilitate the internet of Things (IoT) by linking billions of devices[1, 2]. 3GPP (Third Generation Partnership Project) separated 5G frequency into two portions sub-6 GHz and above 24 GHz or mm-wave frequency. The FCC (Federal Communications Commission) also proposed 24 GHz, 28 GHz, 37-39 GHz, and 60 GHz frequency for mm-wave communications. The (mm-wave) 28 GHz and 38 GHz frequency has excellent characteristics for 5G wireless communication with a wide operating bandwidth. However, mm-wave frequency suffers from attenuation which decreases antenna radiation performance. The high gain antenna is required to overcome the propagation loss at mm-wave frequency[3]. The Wi-Fi frequency-divided into two bands (2.4 and 5 GHz frequency) which has lower propagation loss than mm-wave frequency. The integration of the upper Wi-Fi frequency band and 5G mm-wave frequency will increase the frequency ratio of the operating antenna. The 28 GHz frequency will be appropriate in the urban area for dense 5G small cell networks, where an extra facility like the integration of Wi-Fi frequency will increase the antenna capability. To cover the mm-wave band and Wi-Fi frequency band are important for 5G wireless communication by using a single antenna. Metamaterials are also attractive due to their un-usual property like negative permittivity, permeability, or refractive index, which will provide additional resonance on the antenna[4, 5]. Different research work has been done for designing dual-band antenna for sub-6 GHz and mm-wave frequency, among them only a few has frequency reconfigurable property but suffer for larger size, low gain, or lower efficiency. The frequency reconfigurable property is vital for providing efficient frequency management of the system and reducing antenna power consumption. In[6], a triple frequency band monopole antenna with a franklin strip is presented for sub-6 GHz and mm-wave frequency. An aperture sharing array antenna has been designed for 3.5 GHz and 28 GHz frequencies[7]. Micro-fluidically operated frequency reconfigurable monopole antenna is presented in[8] for sub-6 GHz frequency range for the triple-band application. Dual-polarize frequency reconfigurable dipole antenna has been presented in[9] for sub-6 GHz 5G application with lower efficiency. In [10] a proximity coupling feeding frequency reconfigurable antenna is presented for lower Wi-Fi and sub-6 GHz 5G applications. The designed antenna in[5, 6] has no frequency reconfigurable property, where antenna in [10] provide frequency reconfigurable behavior but suffer from a larger size, lower gain or efficiency also does not integrate mm-wave frequency band which is vital for upcoming 5G application.

This paper proposed a double negative HMR base frequency reconfigurable antenna, which integrates upper Wi-Fi (5 GHz) frequency band and 5G mm-wave (28) frequency band. The frequency reconfigurable property is achieved by using a PIN diode. The operating band of the antenna is 800 MHz (4.5-5.3 GHz) for Wi-Fi frequency and 1020 MHz (27.35-28.37 GHz) for mm-wave frequency, which covers the total allocated frequency for this frequency band

2. Antenna design

2.1. Metamaterial unit cell design

Figure 1(a) shows a hexagonal metamaterial structure (HMR) that has been designed on rogers RT5880 substrate, which thickness is 0.508 mm, dielectric constant 2.2, and loss tangent 0.009. The copper with electrical conductivity 5.96e+07 [S/m] has been used for designing resonation patches. The parameters list of the proposed HMR has been listed in table 1. The simulation setup of the proposed HMR is shown in figure 1(b), where two waveguide port is used along the x-axis. Two C shape hexagonal slot is made for an antiparallel current flow shown in figure 1(c), which facilitate the reverse propagation of EM wave due to negative metamaterial property value (permittivity, permeability, and refractive index). Proposed HMR achieved negative of real and imaginary part for permittivity, permeability, and the refractive index value at 5 GHz frequency band shown in figure 2. All the metamaterial property has been calculated in MATLAB from s-parameters of HMR by using the Nicolson Ross method [4].

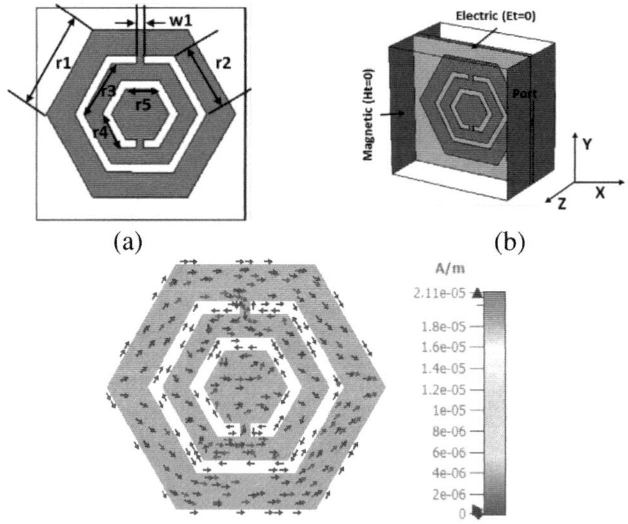

(a) (b)

(c)

Figure 1. (a) Hexagonal metamaterial unit cell, (b) Simulation set-up,(c) Current distribution at 5 GHz frequency.

Table 1. HMR parameters list.

Parameters	Value (mm)
r1	5.0
r2	3.5
r3	3.0
r4	2.0
r5	1.5
w1	0.4

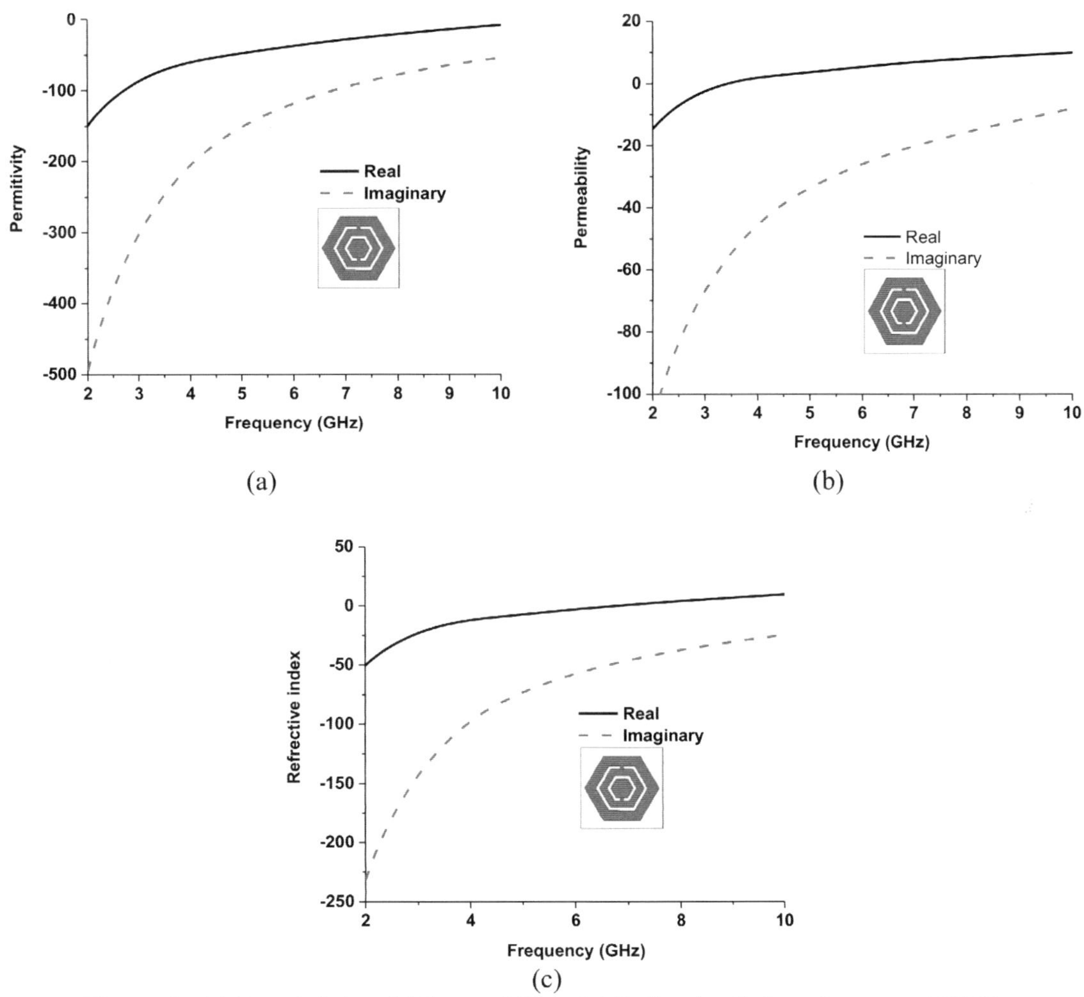

Figure 2. (a) Permittivity, (b) Permeability, (c) Refractive Index of proposed HMR.

2.2. Antenna design and Results analysis

Evaluation of the proposed frequency integration antenna is shown in figure 3. The CST studio software has been used for designing and simulating the proposed antenna. In design 1 a single microstrip patch is designed with an E-shape slot for mm-wave frequency, which resonant frequency is only 27.86 GHz shown in figure 4. Four pair C-shape hexagonal resonating slot is made in-ground plane in design 2, which increases the gain of the proposed antenna. Figure 5 shows the realized gain plot of the proposed antenna with and without resonating slots. A similar S11 curve is attained for both designs 1 and 2 shown in figure 4. In design 3 an HMR is placed on design 2, where HMR is linked with a PIN diode with E- shape slotted patch antenna. For achieving frequency integration PIN diode ON states are applied. The proposed design 3 (final design) realize a single frequency band at the mm-wave frequency (27.86 GHz) for PIN diode OFF state and for PIN diode ON state it will integrate upper Wi-Fi frequency (5 GHz) band, The S11 plot of PIN diode ON and OFF state shown in figure 4. In CST there is no direct PIN diode for ON and OFF state RLC series and parallel circuit are used respectively, were for series circuit R=3Ω, C=0 F, L=2.63e-13 H, and parallel circuit R=20 KΩ, C=0.4e-12 F, L=0 H. The final design of the HMR antenna with all design parameters is presented in figure 6. table 2 listed all values of designed parameters.

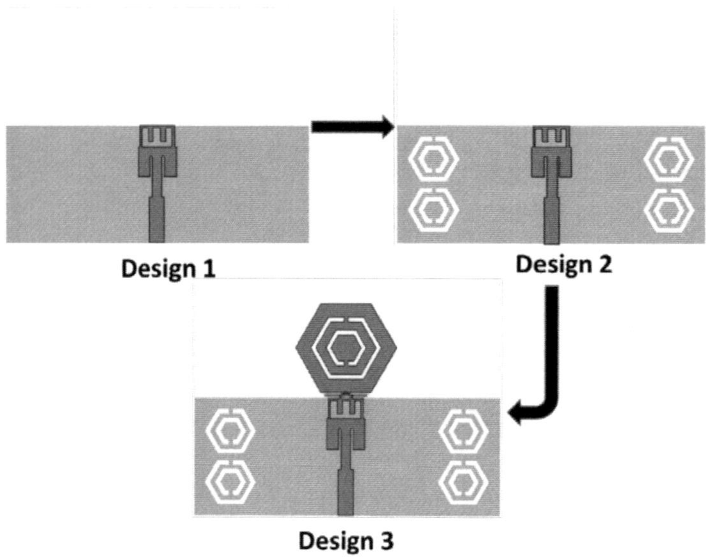

Figure 3. Evaluation of proposed frequency reconfigurable HMR antenna.

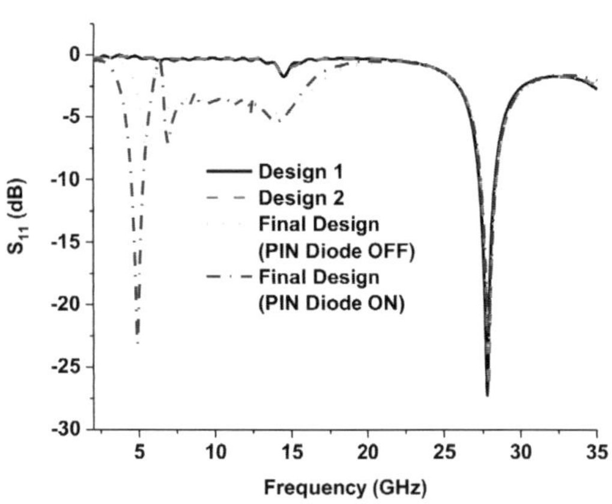

Figure 4. S11 plot of different designed antenna.

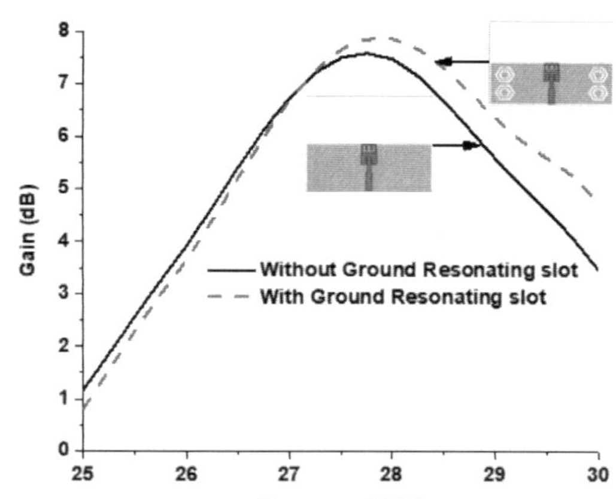

Figure 5. The gain plot of without and with C-shape hexagonal resonating slot on the ground plane.

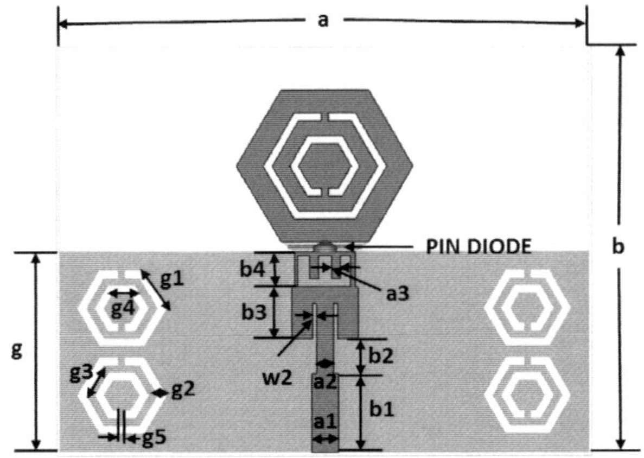

Figure 6. The final design of the proposed HMR antenna.

Table 2. Proposed antenna parameters list.

Parameters	Value (mm)	Parameters	Value (mm)
a1	1.5	g	11.5
a2	1	g1	2.5
a3	0.5	g2	0.5
b1	4.5	g3	2
b2	2	g4	1
b3	2.95	g5	0.4
b4	2.04	w2	0.2

The resonant of the proposed HMR antenna can be understood from the current distribution of both PIN diode ON and OFF states. Figure 7 shows the current distribution for PIN diode OFF state at 5 GHz [figure 7(a)] and [figure 7(b)]. Due to the OFF state current flow from the E-shape slot, MPA to HMR is very less, which is unable to resonating the HMR and does not realize any frequency band at 5 GHz. However, E-shape slot MPA shows high current density at 27.86 GHz frequency for OFF state, which attained resonant frequency at mm-wave frequency. The C shape hexagonal resonating slot creates an anti-parallel current flow on the ground shown in figure 7(c), which provides additional resonant for gain enhancement at mm-wave frequency. The frequency integration of the 5 GHz Wi-Fi band can be understood from the PIN diode ON state. The HMR fully resonates at 5 GHz frequency due to the maximum current flow. The mm-wave frequency band (27.86 GHz) is also attained in the ON state. Though at 27.86 GHz current flow to HMR is reduced and the maximum current is concentrated on E-shape slot MPA shown in figure 8.

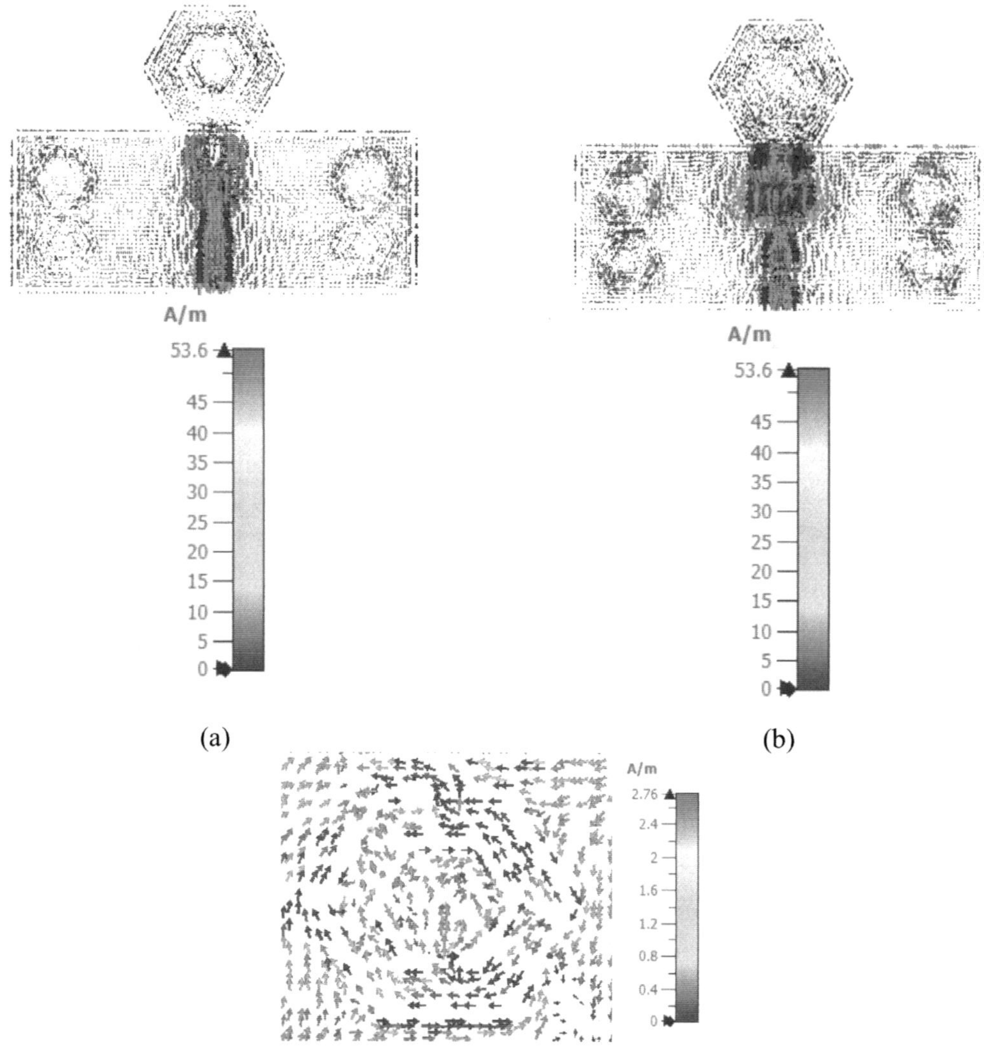

(c)

Figure 7. Current distribution of PIN diode OFF state (a) 5GHz and (b) 27.86 GHz, (c) Hexagonal resonating slot in ground.

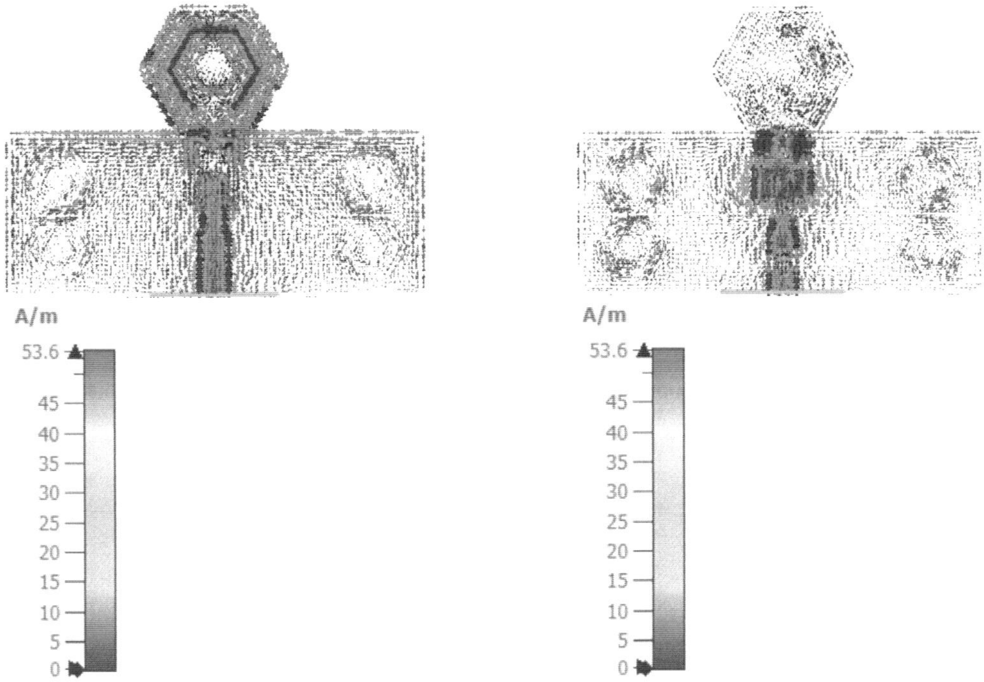

(a) (b)

Figure 8. The current distribution of PIN diode ON state (a) 5GHz and (b) 27.86 GHz.

The 3-dimensional (3-D) and 2-dimensional (2-D) far-field radiation patterns are presented in figures 9 and 10 respectively. An omnidirectional radiation pattern attained at 5 GHz frequency with 2.38 dBi gain because there has no ground plane backside of the HMR, which provides a 5 GHz resonant frequency. On the other side at the mm-wave frequency, 27.86 GHz resonant frequency is achieved due to the E-shape slot MP. Four pair C-shape hexagonal resonating slot prevents backpropagation of antenna radiation and provides directional far-field radiation patterns with peak 7.12 dBi gain. The gain and efficiency plot of the proposed antenna is presented in figure 11. It is shown from figure 11 that the antenna depicts 2.38 dBi and 7.12 dBi gain at5 GHz and 27.86 GHz frequency, respectively. Moreover, the total efficiency achieved at 5 GHz and 27.86 GHz frequency are is 86% and 90% respectively.

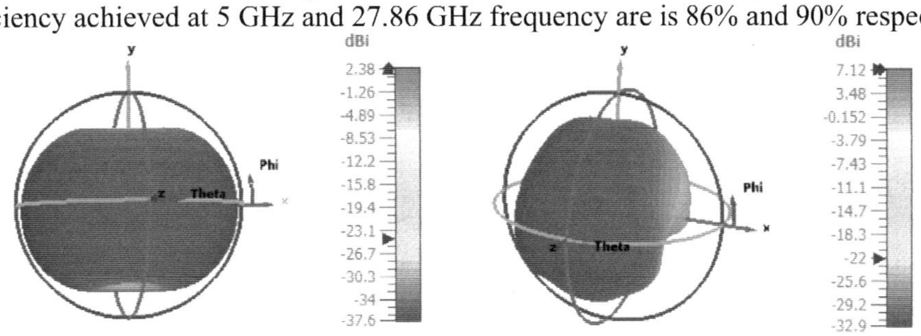

(a) (b)

Figure 9. 3-Dimensional far-field radiation pattern at (a) 5GHz and (b) 27.86 GHz.

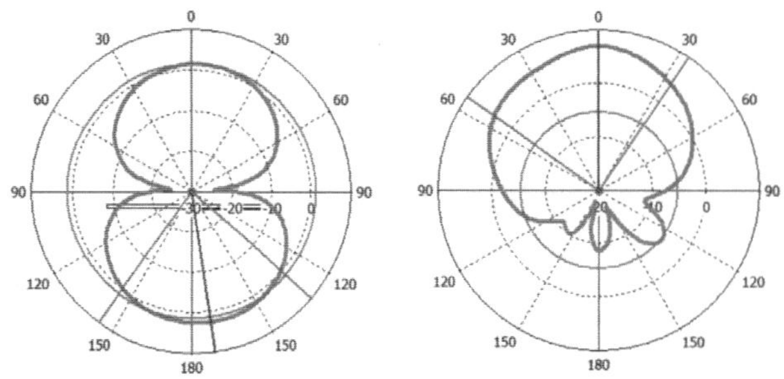

(a) (b)

Figure 10. 2-Dimensional far-field radiation pattern at (a) 5GHz and (b) 27.86 GHz.

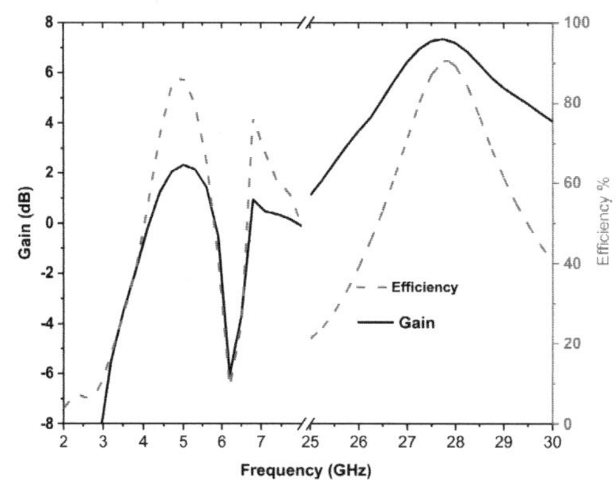

Figure 11. Gain and efficiency polt of proposed HMR antenna

3. Comparison

A detailed comparison of the proposed HMR antenna with recent work is presented in table 3. It is shown from the table 3 that the antenna [6] and [7] attained sub-6 GHz frequency and mm-wave frequency band simultaneously, but does not provide frequency reconfigurable property. Besides, the antennas in [8-10] provide frequency reconfigurable facilities. However, theses antennas do not cover the mm-wave frequency band. The proposed frequency reconfigurable antenna covers Wi-Fi and mm-wave frequency simultaneously with significant gain and efficiency.

Table 3. Comparison with existing work

Reference	Size	frequency	Gain (dBi)	efficiency	Frequency Reconfigurable?
[6]	40×40×0.508	2.4/5.5/28	1.9/3.8/7.4	--/--/--/	No
[7]	43×33.9×0.245	3.5/28	7.07/11.31	N/A	No
[8]	90×45×3.5	2/2.6/3	2.7/3/3.1	92/89/87	Yes
[9]	72×72×3.5	3.5/5.5	6.86/8.14	70/80	Yes
[10]	100×100×5.7	2.45/3.5	6.5/6.8	64.5/69.5	Yes
Proposed	30×23×0.508	5/27.86	2.38/7.12	86/90	Yes

4. Conclusion

This paper proposed frequency integration of Wi-Fi frequency (5 GHz) and 5G mm-wave frequency by frequency reconfigurable techniques. A PIN diode has been used to link the HMR and E-shape slot MPA. PIN diode OFF sate provides a single frequency at 27.86 GHz mm-wave frequency, its ON state integrates the Wi-Fi frequency (5 GHz) band. The operating bandwidth is 800 MHz (4.5-5.3 GHz) and 1020 MHz (27.35-28.37 GHz) at 5 GHz and 27.86 GHz resonant frequency respectively. the peak gain and efficiency of the designed antenna are 2.33, 7.12 dBi, and 86%, 90% at 5 GHz and 27.86 GHz frequency individually.

5. References

[1] Palattella M R, Dohler M, Grieco A, Rizzo G, Torsner J, Engel T and Ladid L 2016 Internet of things in the 5G era: Enablers, architecture, and business models *IEEE journal on selected areas in communications* **34** 510-27

[2] Rappaport T S, Sun S, Mayzus R, Zhao H, Azar Y, Wang K, Wong G N, Schulz J K, Samimi M and Gutierrez F 2013 Millimeter wave mobile communications for 5G cellular: It will work! *IEEE access* **1** 335-49

[3] Sulyman A I, Nassar A T, Samimi M K, MacCartney G R, Rappaport T S and Alsanie A 2014 Radio propagation path loss models for 5G cellular networks in the 28 GHz and 38 GHz millimeter-wave bands *IEEE communications magazine* **52** 78-86

[4] Hakim M L, Alam T, Almutairi A F, Mansor M F and Islam M T 2021 Polarization insensitivity characterization of dual-band perfect metamaterial absorber for K band sensing applications *Scientific reports* **11** 1-14

[5] Mohammad Lutful Hakim T A, Mohamed S. Soliman, Sami H. A. Almalki and Mohammad Tariqul Islam 2022 Polarization Insensitive Symmetrical Structured Double Negative (DNG) Metamaterial Absorber for Ku-band Sensing Applications *Scientific reports*

[6] Yassin M E, Mohamed H A, Abdallah E A and El-Hennawy H S 2019 Single-fed 4G/5G multiband 2.4/5.5/28 GHz antenna *IET Microwaves, Antennas & Propagation* **13** 286-90

[7] Lan J, Yu Z, Zhou J and Hong W 2019 An aperture-sharing array for (3.5, 28) GHz terminals with steerable beam in millimeter-wave band *IEEE Transactions on Antennas and Propagation* **68** 4114-9

[8] Dey A and Mumcu G 2015 Microfluidically controlled frequency-tunable monopole antenna for high-power applications *IEEE Antennas and Wireless Propagation Letters* **15** 226-9

[9] Nie Z, Zhai H, Liu L, Li J, Hu D and Shi J 2019 A dual-polarized frequency-reconfigurable low-profile antenna with harmonic suppression for 5G application *IEEE Antennas and Wireless Propagation Letters* **18** 1228-32

[10] Jin G, Deng C, Yang J, Xu Y and Liao S 2019 A new differentially-fed frequency reconfigurable antenna for WLAN and Sub-6GHz 5G applications *IEEE Access* **7** 56539-46.

Acknowledgments

This work was funded by the research grant no: FRGS/1/2020/TK0/UKM/03/1.

Near field radio link design in fat phantom for Wireless Capsule Endoscopy application

Nur Amalina Kamaruddin[1*], Kamilia Kamardin[1], Yoshihide Yamada[1], Norsiha Zainudin[2], Takuji Arima[3], Naobumi Michishita[4], and Masaharu Takahashi[5]

[1] Communication Systems and Network iKohza, Department of Electronic Systems Engineering, Malaysia-Japan International Institute of Technology, Universiti Teknologi Malaysia, Kuala Lumpur 54100, Malaysia
[2] Department of Electrical Engineering, Faculty of Engineering, University of Malaya, Kuala Lumpur 50603, Malaysia
[3] Dept. of Electronic and Information Engineering, Tokyo University of Agriculture and Technology, Koganei, Tokyo 184–8588, Japan
[4] National Defense Academy, Yokosuka, Japan
[5] Center for Frontier Medical Engineering, Chiba University, Chiba, Japan
[*] nuramalinakamaruddin@graduate.utm

Abstract. Medical devices such as Wireless Capsule Endoscopy (WCE) that need wireless imaging and monitoring, operating at 402 MHz, is considered to be in the near field region due to the distance being shorter than the wavelength. In this paper, the radio link calculation method in the near field region is investigated. Normal Mode Helical Antenna (NMHA) is used as the transmitter antenna which is placed inside of a fat phantom of 11.6 F/m and the receiver antenna at distance 0.1 to 1.0 m apart. For link budget establishment, parameters such as power density, W and receiver antenna effective area, A_e are investigated through electromagnetic simulation. The distance dependency of W is examined and the result of $W \propto 1/(\text{distance})^2$ is obtained. The relation of A_e of the receiver antenna is found to be the size of the cross-section area of the antenna. A radio link equation is formed by employing W, A_e and other propagation loss factors of a medium and is then compared with the power received, P_R and S_{21} obtained from simulation. The agreement between calculated and simulated results shows that this equation can be used in the near field region.

1. Introduction

Wireless Capsule Endoscopy (WCE) is one of the medical devices that has the purpose of diagnosing the health conditions inside the gastrointestinal tract. Evaluation of link budgets in the near field by [1] and [2] based on Friis equation and numerical evaluation using FEKO shows that the Friis equation is possible to use for the near field with some adjustment on the original equation but the research is based on the antenna in free space, not in the human body. Thus, link budget equation for near field and in the human body is desirable.

In this paper, analysis on power density and equivalent area at receiver antenna are analysed through electromagnetic simulation. The analysis is done in the near field region to check its adaptability. By using FEKO simulator software, the theoretical understanding about signal propagation can be compared with the simulation results to prove the hypothesis. Establishment of a suitable link design equation for WCE system in the near field condition is important for precise design of WCE system. This is due to most of the application of WCE is in near field regions with low operation frequency. The derivation of the link design equation could improve the performance of WCE and improve the knowledge in signal propagation in the near field region.

2. Near field link budget equation

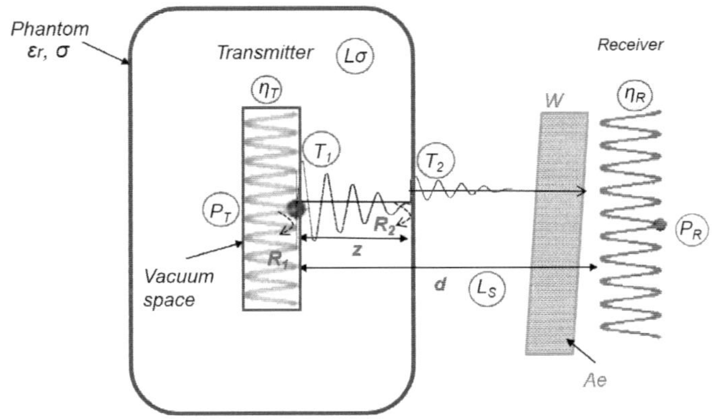

Figure 1. Signal transmission from transmitter to receiver

Figure 1 shows the illustration of signal transmission from NMHA inside a human body phantom to a receiver outside the phantom separated by distance, *d*. Based on the parameters deduced from figure 1, propagation link for transmission in the near field is proposed as equation (1). Based on equation (1), the parameters are received power (P_R), transmitted power (P_T), efficiency of the transmitter (η_T) and receiver antenna (η_R), transmitted signal from capsule to phantom (T_1), transmitted signal from phantom to free space (T_2), dielectric loss (L_σ), spherical spreading of power (L_S), and effective antenna aperture (A_e). Table 1 shows the descriptions of the parameters in equation (1).

$$P_R = P_T + \eta_T + T_1 + L_\sigma + T_2 + L_S + A_e + \eta_R \tag{1}$$

Table 1. Equations for link budget

Link Budget	Description of parameters	Expressed by equation	
Transmitted power, P_T	Power transmitted	20 dBm (0.1 W)	
Antenna efficiency, η	R_a, Radiation resistance R_{in}, Antenna input resistance	$\eta = 10\log\frac{R_a}{R_{in}}$	(2)
Transmission coefficient at the boundary, T	n, Refractive index of the phantom	$T = 10\log\frac{4n}{(n+1)^2}$	(3)
Propagation loss in phantom, L_σ	ε, Dielectric constant σ, Conductivity ω, Angular frequency α, Attenuation constant z, radius of phantom	$L_\sigma = 10\log e^{-2\alpha z}$ $\alpha = \sqrt{\frac{\omega^2\varepsilon\mu_0}{2}}\left(\sqrt{1+\left(\frac{\sigma}{\omega\varepsilon}\right)^2}-1\right)^{\frac{1}{2}}$	(4) (5)
Power spreading in free space, L_S	d, Distance to the observation point	$L_S = 10\log\frac{1}{4\pi d^2}$	(6)
Effective area of the receiving antenna, A_e	W, Power density η_R, Efficiency of receiver antenna	$A_e = 10\log\left(\frac{P_R}{W\eta_R}\right)$	(7)

Efficiency of transmitter antenna, η is shown as equation (2) where R_a is the radiation resistance and R_{in} is the input impedance of the antenna. R_{in} is the pure resistance when NMHA reaches a self-resonant structure [3]. When signals propagate through one medium to another, reflection loss occurs. Parts of the signal waves are reflected back while some of it will be transmitted when signal travelled into a different material [4]. The relationship of transmission coefficient, T and reflected signal, R is shown as equation (3) where for T_1, $n = \sqrt{\varepsilon_r}$ and for T_2, $n = 1/\sqrt{\varepsilon_r}$. In this paper, the transmitted signals are considered to reflect twice; first from vacuum (inside the capsule) to the phantom layer, second from phantom layer to free space.

Dielectric loss can be calculated by using equation (4) where the thickness of dielectric medium, z and attenuation constant, α can be calculated using equation (5). The permeability of a biomedical materials, $\mu_r = 1$. Thus, μ is equal to μ_0. Based on the definition of power density at receiver, P_R can also be expressed as equation (6) where power density is W, antenna equivalent area is A_e and receiver antenna efficiency is η_R. From free space loss factor in Friis transmission equation, spherical spread of power exists to show the loss when the transmission distance increases. This equation can is shown in equation (8) [5].

Although theoretically equation (6) until equation (8) were mentioned by [5–8], analytical results to confirm these equations are not yet tested for NMHA in the human body for near field propagation. Clarification of L_S and A_e will strengthen the agreement of the proposed equation to be used to estimate link budget of NMHA in the human body.

3. Simulation of NMHA in FEKO

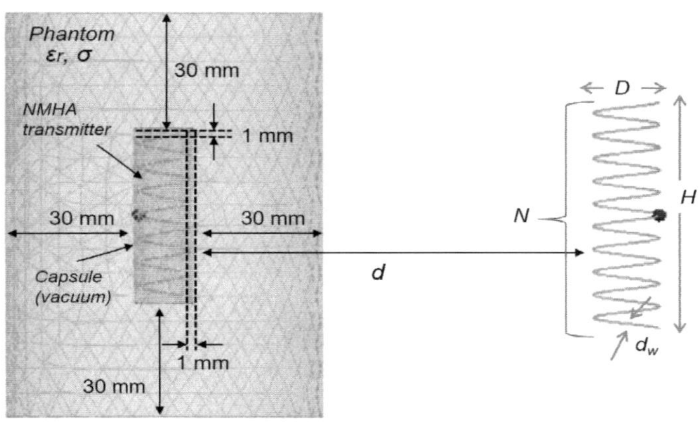

Figure 2. Model of NMHA in FEKO

Table 2. Parameter of transmitter and receiver antenna

Parameters	Transmitter	Receiver
Frequency	402 MHz (λ = 0.75 m)	
Material	Copper	
Wire conductivity, σ_w	58x10^6 [1/Ωm]	
Diameter of wire, d_w	1.2 [mm]	
Condition	Fat phantom	
Dielectric constant, ε_r	11.6 F/m	Free space
Conductivity, σ	0.1 S/m	
Height, H (mm)	43.8	59.7
Diameter, D (mm)	10.9	15.3
No. of turns, N	7	10
R_{in} (Ω)	25.0	2.99
R_a (Ω)	0.33	2.63
Efficiency, η (dB)	-18.76	- 0.56

A model of NMHA inside a human body phantom with a receiver antenna is simulated by using FEKO. The parameters for NMHA inside of human body tissues phantom model are shown in Figure 2. The size of the phantom is 30 mm from the transmitter antenna's dimension. Another NMHA as the receiver is added to the model. The distance between transmitter and receiver antenna is separated by the distance, *d*. The parameters for simulation are tabulated in Table 1. Diameter, height, number of turns and wire diameter are expressed as *D*, *H*, *N* and d_w respectively as shown in figure 2. Both transmitter and receiver antenna are designed according to self-resonant structure to ensure their efficiency. No polarization mismatch or impedance mismatch for both antennas.

4. Derivation of link design equation in fat phantom

4.1 Power spreading factor

Simulation software, FEKO is used to measure the value of *W* at each distance at 0.1 to 1.0 m from the transmitter. Power density is viewed in the near field distribution in order to check its pattern and magnitude when the distance of measured point increases. Figure 3 shows the results of power density at the distance of 0.1 m. The near field distribution results from figure 3 can be seen that the shape of the power density spreading is in spherical shape.

Power density, *W* can also be expressed in equation (8) where power transmitted is divided with the surface of a sphere. As shown in equation (8), the inverse surface of a sphere is considered as a power spreading loss factor. Power transmitted is constant for all conditions. The relationship between *W* and L_S is compiled in figure 4. Based on figure 4, the degradation pattern of *W* and L_S are the same. The results obtained for power density shows that power density decreases when the distance between transmitter and receiver increases. From the near field distribution, it can be seen that the power density spreads in spherical shape. Spherical spreading loss has the same degradation pattern as power density and come with the conclusion that $W \propto 1/(distance)^2$.

$$W = \frac{P_T}{4\pi d^2} \tag{8}$$

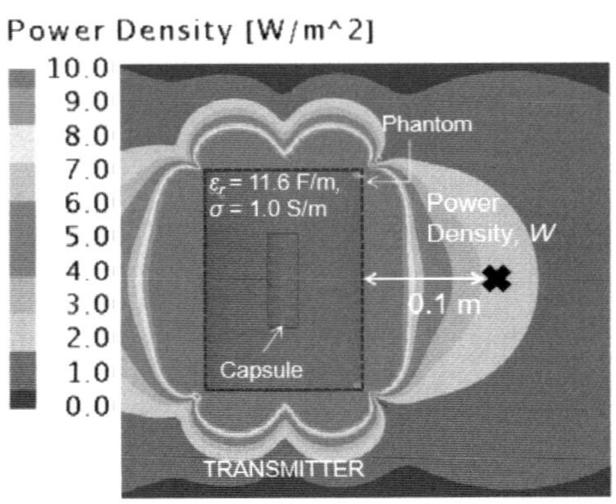

Figure 3. Power density separated by a distance of 0.3 m

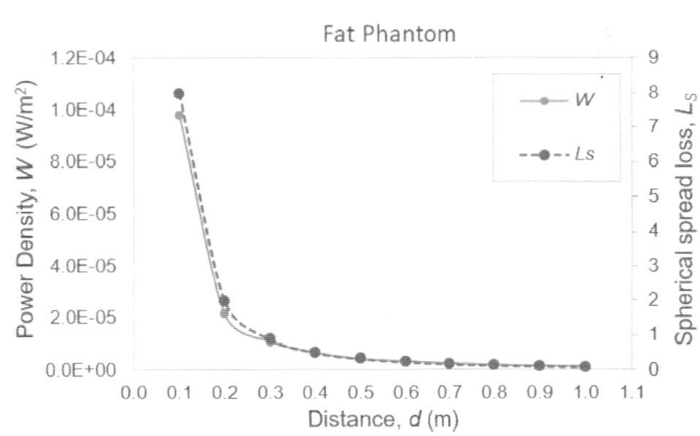

Figure 4. Power density and spherical spreading loss at different distance

4.2 Effective aperture of receiver antenna

Effective area, A_e can be calculated as equation (7). Figure 5 shows the comparison of effective aperture of receiver antenna with the structure of the receiver antenna. The data shown in figure 5 is the average value of A_e for distance 0.1 to 1.0 m. This shows that A_e has the value of 9.17 cm^2 which is almost equivalent to the surface dimension of the receiver antenna and no mismatching occurred here. Figure 6 shows the graph of calculated A_e using equation (9) with the value obtained from P_R and *W*. Based on Figure 6, the values of A_e is spreading near to the value of A_P which is the surface dimension of the receiver antenna. From the results show that A_e is almost as same as A_P. This leads to the conclusion where A_e can be calculated by using equation (7).

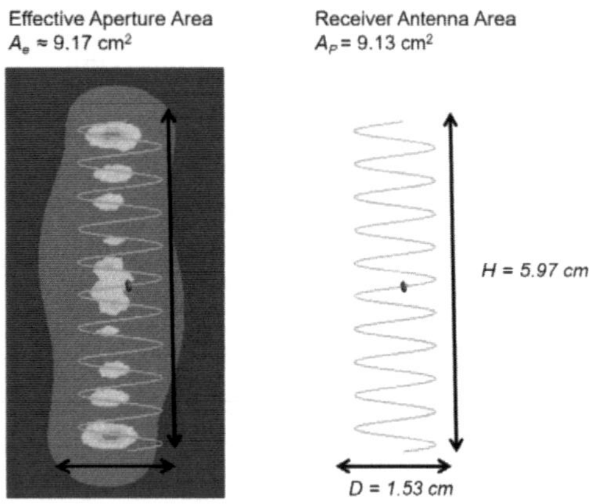

Figure 5. Comparison of effective aperture and antenna structure

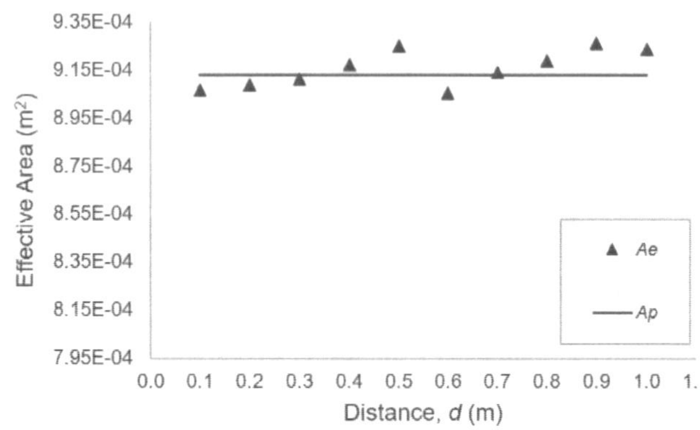

Figure 6. Effective area comparison for fat phantom

4.3 Comparison of Calculated S_{21} with Simulated S_{21}

In order to obtain the value of calculated S_{21}, P_R is divided with P_T as shown in equation (9). The results obtained from calculation using equation (9) are then compared with the simulation results obtained in FEKO software to ensure the agreement between them.

$$S_{21} = \frac{P_R}{P_T} \tag{9}$$

By using equation (1) and equation (9), power received and S_{21} for fat is calculated. The values of calculated parameters are tabulated in table 3. From table 3 P_T is kept constant at 20 dBm and other parameters such as η_T, η_R, T_1, T_2, L_σ and A_e are also constants. A_e is considered constant because the same receiver antenna is used throughout the distance until 1.0 m. The parameters that are affected by the increase in distance from the transmitter are L_S, P_R and S_{21} only.

Table 3. Calculated Parameters for Fat Phantom

P_T (dBm)	η_T (dB)	T_1 (dB)	L_σ (dB)	T_2 (dB)	A_e (dB)	η_R (dB)
20	-18.8	-1.54	-9.12	-1.54	-30.3	-0.56

d (m)	0.1	0.2	0.3	0.4	0.5	0.6	0.7	0.8	0.9	1.0
L_S (dB)	9.01	2.99	-0.53	-3.03	-4.97	-6.56	-7.89	-9.05	-10.08	-10.99
P_R (dBm)	-33.6	-39.8	-43.3	-45.8	-47.4	-49.4	-50.6	-51.8	-52.7	-53.9
S_{21} (dB)	-53.6	-59.8	-63.3	-65.8	-67.4	-69.4	-70.6	-71.8	-72.7	-73.9

From simulation, the results that can be obtained are P_R and S_{21}. The result for comparison between calculated S_{21} and simulated S_{21} is shown in figure 7. The graph in figure 7 shows that the equation used to calculate S_{21} of NMHA in fat phantom agree well with the simulation results obtained in FEKO software. This shows that the equation proposed for estimating radio link budget for NMHA in human body phantom is possible to be used for practice.

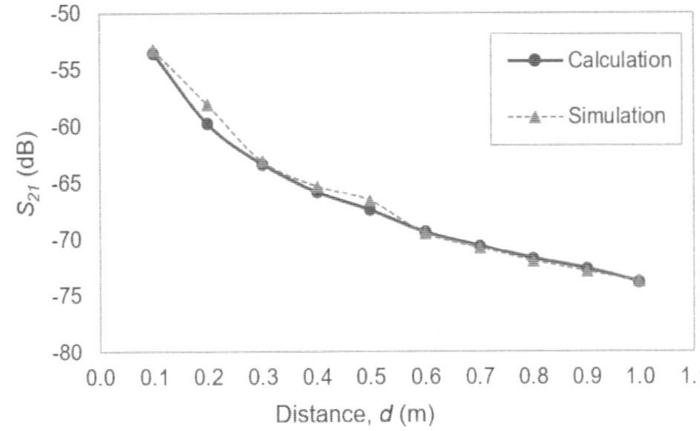

Figure 7. Comparison of calculated S_{21} with simulated S_{21}

5. Conclusion

The important parameters for radio link design in near field region are the power spreading in free space and the receiver antenna's received area which was employed for the NMHA. By the electric power density results calculated through the electromagnetic simulator, important parameters have been estimated. Through the results, it is clarified that the power spreading factor decreased by the square value of the distance. In addition, the results of the antenna equivalent area of the receiver antenna are almost the same as the cross-sectional area of the receiver antenna. The derivation of the radio link budget equation is completed after the analysis on the power spreading factor and antenna equivalent area were done. In order to prove the derivation of the proposed equation as an equation for estimation of link budget for NMHA, the calculated P_R and S_{21} is compared with simulation. From the comparison, it shows that the proposed equation can be used to estimate radio link budget for NMHA inside of the human body phantom.

6. References

[1] Kim I, Xu S and Rahmat-Samii Y Generalised correction to the Friis formula: quick determination of the coupling in the Fresnel region 2013 *IET Microwaves, Antennas & Propagation* **7** 1092
[2] Schantz HG, Unden AH, Nikravan MA, Kwon DH Simple formulas for near-field transmission gain and fields. 2013 *Antenna Applications Symposium*
[3] Kamaruddin NA, Kamardin K and Yamada Y Electric Field Distributions of NMHA Inside and Outside of a Human Body Phantom 2020 *IEEE International RF and Microwave Conference (RFM)* (pp. 1-4)
[4] Cheng DK 1986 *Field and Wave Electromagnetics* Vol. 28, IEEE Antennas and Propagation Society Newsletter. Addison-Wesley p 27–8
[5] Nikolova NK, Ravan M and Amineh RK 2012 *Substrate-Integrated Antennas on Silicon* Advances in Imaging and Electron Physics
[6] Stutzman WL and Thiele GA 2012 *Antenna theory and design*. John Wiley & Sons
[7] Balanis CA 2016 *Antenna Theory and Design*. 4th Edition. John Wiley & Sons
[8] Cerwin S 2019 *Radio Propagation and Antennas: A non-mathematical treatment of radio and antennas* AuthorHouse

Acknowledgments

Authors wishing to acknowledge financial support from the Ministry of Higher Education (MOHE) through Fundamental Research Scheme (FRGS/1/2018/TK04/UTM/02/16).

Journal of Physics: Conference Series 2250 (2022) 012003 doi:10.1088/1742-6596/2250/1/012003

LTE-based passive forward scattering radar for drone detection and classification

N H Abdul Aziz[1*], M H Mohd Fodzi[1], K K Mohd Shariff[2] and M A Haron[1]

[1]Antenna Research Centre, School of Electrical Engineering, College of Engineering, Universiti Teknologi MARA, 40450 Shah Alam, Selangor, Malaysia.
[2] Microwave Research Institute, Universiti Teknologi MARA, 40450 Shah Alam, Selangor, Malaysia.

Email: noor4083@uitm.edu.my

Abstract. Passive forward scattering radar is a system that consists of a transmitter that generates microwaves domain, and at the receiver part, receives LTE-based signal from the transmitter, which will result in the Doppler signature of any object. This system can actually be used for detecting the presence of guided missiles, human, aircraft and drone that pass through between the transmitter and receiver. For this study, the main objective is to detect a drone, which are mainly used by humans to trespass or enter secure and private areas. Recently, there have been a few cases of burglary in several states where they have made use of drones to monitor the owner of the house before entering that house. For that reason, this study can help to increase the security and protection against drone trespassing. Specifically, the detection of a drone was carried out by determining the Doppler signature of the drone at two different heights, which are 2 meters (lower) and 3 meters (upper) from the ground by using passive scattering radar. All of the experimental data collected from this study were processed using the MATLAB software to compare the result of drone detection at different heights through Principal Component Analysis (PCA) results. This analysis, from the upper and lower heights, shows a good detection and classification of the drone, however lower height indicated better performance in PCA and confusion matrices analysis. This is due to the bigger cross-sectional area for lower height compared to the upper height of drone that passed through the forward scatter region. This is useful for border protection and monitoring.

1. Introduction

Radar is an object detection system that specifies the angle, velocity or range of objects using radio waves. It can be used to detect aircraft, ships, spacecraft, guided missiles, motor vehicles, weather formations, and terrain [1]. In this study, to perform drone detection, Forward Scattering Radar (FSR) system was used. FSR is a special mode of bistatic radar that can be used for target detection and classification [2].

Radar transmits radio waves or microwaves that reflect from any object in their path. A receiver radar, which is typically the same system as the transmit radar, receives and processes these reflected waves to determine properties of the object. Passive radar is defined as a system that has a receiver antenna only. The passive radar does not have a transmitter and this system will use existing commercial telecommunication antenna as the transmitter and use its own receiver antenna to receive the radio waves. For this study, a passive FSR system was used to detect the moving target, which is a flying drone, at different heights. Recently, there were a few cases of burglary that involved the use of a drone, which happened in five states, namely Selangor, Kuala Lumpur, Perak, Johor and Seremban

[3]. This situation is worrying because it is happening in Malaysia. This study can help the police to eradicate this activity. The signature of the moving drone can be captured as the Doppler signature. When a moving object infiltrates a radar system, there would be some differences in the signal amplitude and frequency. The differences in the signal are caused by scattering of the signal that happens when the signal collides with the object. The factors that affect the signal are the speed and the size of the object, and also the distance between the object to the receiver. This phenomenon is called the Doppler Effect [4]. Furthermore, passive FSR provides a lot of interesting features such as the generally simple hardware, an improved target radar cross section, a long coherent interval of the receiving signal, protection from stealth technology and possible operation using non-cooperative transmitter [5], [6].

All the process of data captured until processing and analyzing the data was performed using MATLAB software. The principle component analysis was done to prove drone detection in the area. The principle component analysis is to decrease the dimensionality of a data collection, which comprises of a substantial number of interrelated variables [7].

2. Methodology

2.1. Drone as a Moving Target with Different Heights

In this project, a flying drone was used as the moving target, and its details are as in table 1. There were two different heights used as the manipulated parameter for this project. For the upper position, the height was 2 meters from the ground, and for the lower position, it was 3 meters from the ground.

Table 2 explains the number of samples collected at different sites and frequency. The 4G Long Term Evolution (LTE) signals of 1.8 GHz and 2.6 GHz spectrum band are chosen due to transmitting signal from the telecommunications sector and it was assigned by the Malaysian Communications and Multimedia Commission (MCMC). In addition, the LTE frequency of 2.6 GHz is suitable for covering densely populated areas and handling large volumes while the 1.8 GHz could provide two times wider coverage compared to 2.6 GHz and better indoor signal strength.

The difference in number of samples taken at each place and frequency spectrum band due to the different condition and environment that differed from day to day as it can be sunny, cloudy, windy, rainy or stormy, which is the typical climate in tropical Malaysia that give impact in data collection.

Table 1. Information about the drone.

Silhouette	Height (m)	Length (m)	Dimension (m^2)	Material	Weight (g)	Battery power (mAh)
	0.193	0.35	0.06755	ABS (Acrylonitrile butadiene styrene)	645	2000

Table 2. Information about the number of samples.

Location	Number of Samples			
	1.8 GHz		2.6 GHz	
	Upper	Lower	Upper	Lower
Taman Suria	20	18	21	19
Teluk Kemang	18	19	-	-

2.2. Experiment Site

In order to carry out this experiment, two different sites are make sure were suitable and satisfactory. There were several criteria and characteristics that needed to be underlined. For example, the strength of signal, the distance between transmitter and receiver, the blockage of signal and environment safety. For the first experiment, it was done at Taman Suria, UiTM Shah Alam, Malaysia (3.073°, 101.497°). At a distance of 187.5 meters away from the radar receiver, the telecommunication transmitter antenna transmits Long-Term Evolution (LTE) signal with frequency of 1.8 GHz and 2.6 GHz.

Figure 1 shows the location of the experiment as mentioned above. To determine the power of the spectrum of known and unknown signal, a provider can be measured using the spectrum analyzer. As shown in figure 2, the spectrum analyzer reading for the frequency 1.8 GHz was -60 dB for the power of the signal. While figure 3 shows the reading of power of the signal for 2.6 GHz, which is -50 dB.

The second experiment site chosen was by the seaside of Teluk Kemang, Port Dickson, Negeri Sembilan (2.448°, 101.855°), as shown in the figure 4. The distance of the telecommunication transmitter antenna from the radar receiver system was about 580 meters. For this site, there was only 1.8 GHz signal detected based on the founding of spectrum analyzer. Before getting the best signal received from the transmitter tower, some adjustment was needed at the angle and position of the receiver antenna. The accurate position and angle will result in the strongest signal from the transmitter. Figure 5 shows the signal strength of 1.8 GHz. In figure 5, the signal is 1.8 GHz with the power for the signal being -35 dB. Figure 6 shows the arrangement of baseline experiment for the drone detection, which flew between the transmitter and receiver.

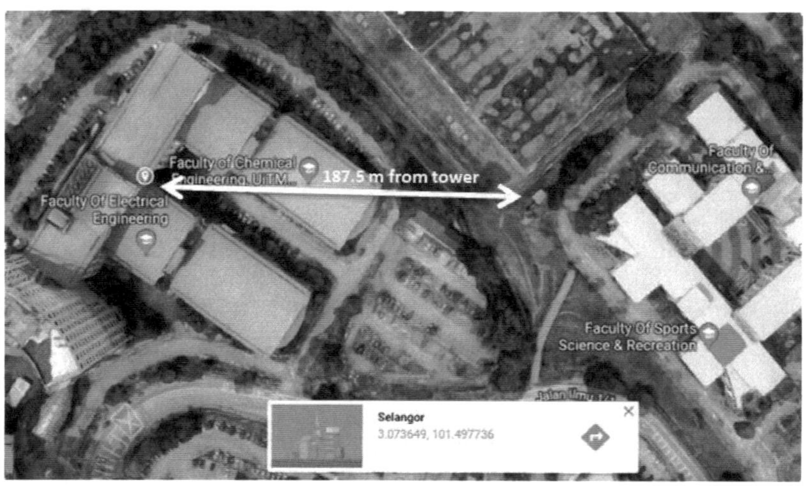

Figure 1. Experiment Site at Taman Suria.

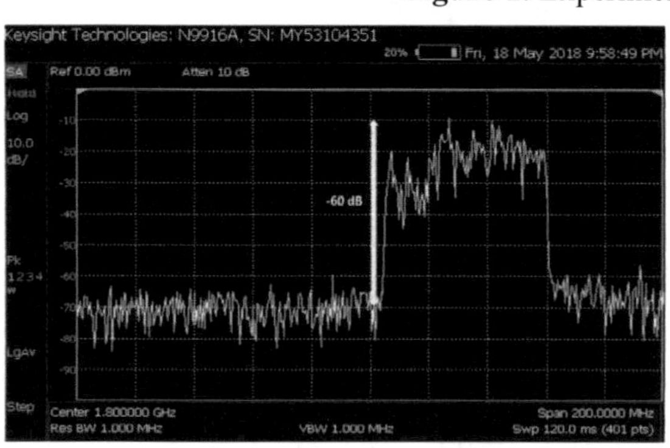

Figure 2. Signal strength for 1.8 GHz at Taman Suria.

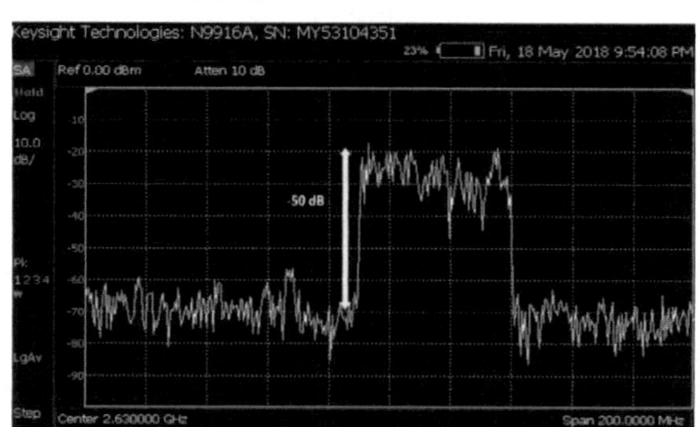

Figure 3. Signal strength for 2.6 GHz at Taman Suria.

Figure 4. Experiment Site at Teluk Kemang.

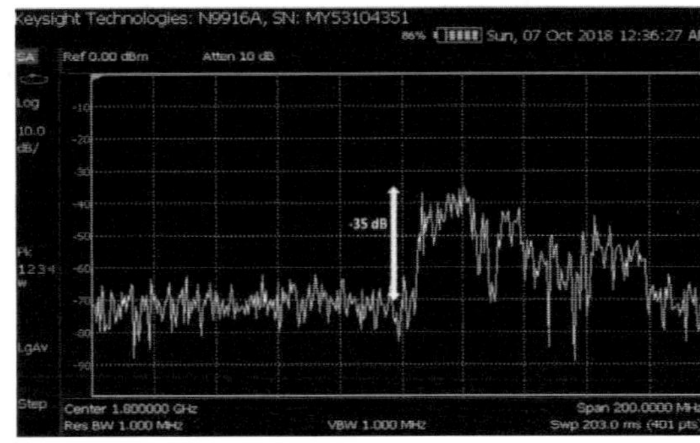

Figure 5. Signal strength for 1.8 GHz at Teluk Kemang.

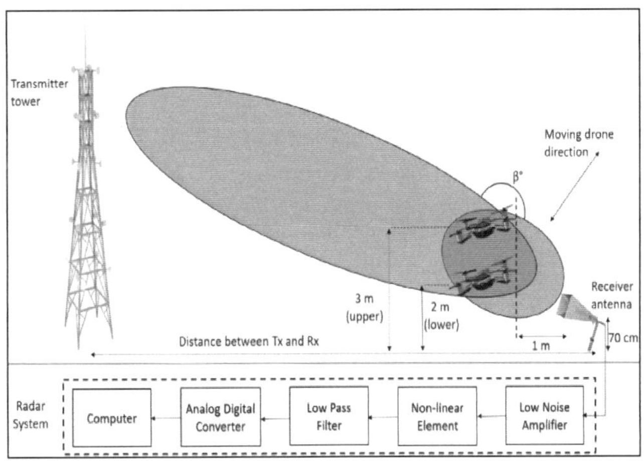

Figure 6. Setup of LTE-Based PFSR for drone detection and classification.

2.3. Signal Processing

For analyzing the process, all of the data that was collected during the experiment was analyzed and processed using a computer software, which is the MATLAB software. The reason was that this software has a high-level of technical ability in analyzing and processing raw data to become an organized visual result. There are a few steps that need to be done when using the MATLAB software. There is a different method to obtain different result that will come out in graph display. In addition, by using the MATLAB software, it also enables the results of the wave pattern and signal of moving drone to be classified.

Referring to figure 7, these are all of the steps for analyzing the data using MATLAB software. To collect the data, the time taken for the Doppler signature was set to 20 seconds. During that duration of time, it will capture and analyze the moving target signature, the flying drone, at different heights. In data segmentation, it was set to be 3 seconds for the Doppler to capture and analyze the flying drone signature. Then, the raw data will go to the denoise process. In this stage, all of the unwanted signals will be denoised using the coding in the MATLAB software, leaving almost perfect signature of the drone only. After that, all of the denoised data from the denoise signal process will be transformed in the Power Spectral Density (PSD) frequency domain. Lastly, the Principal Component Analysis (PCA) of frequency domain will represent some of the values in the scatter plot. In the end, in the classification process, the different height of flying drone signature will be classified into their own cluster.

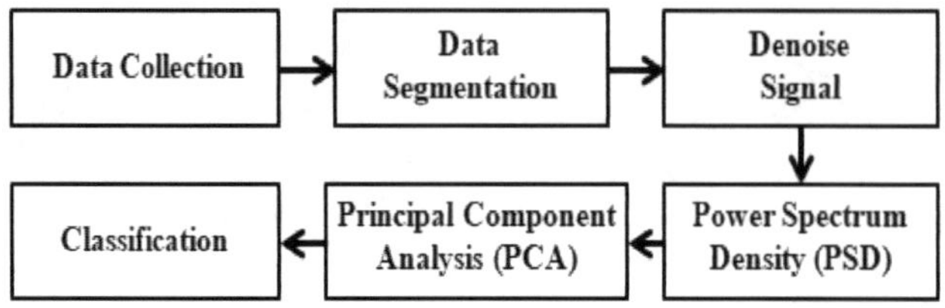

Figure 7. The flowchart of data analysis process.

3. Result And Discussion

3.1. Time Domain After Denoise Process

Based on the results, after denoising the unwanted signal, a time domain signal was needed to represent the Doppler signature of different heights of the flying drone. Due to the different frequency in Teluk Kemang and Taman Suria, there were different Doppler signatures. However, the MATLAB software was unable to identify the Doppler signature only, so the denoise process did not remove all of the unwanted signal in the signal obtained.

Figure 8, figure 9 and figure 10 show the result of the denoised signature of the flying drone at different heights. The signatures were different because of the different frequencies and different locations, even though the same drone was used in these experiments. The red line represents the upper position and the blue represents the lower position of the flying drone.

Figure 8 explains the time domain of the flying drone at different heights with the frequency of 1.8 GHz at Taman Suria. For the upper position, the drone took about 1.8 seconds to pass through between the transmitter and receiver compared to the lower position with 2.3 seconds. There was a smaller difference on the time taken for the drone flying, which was 0.1 seconds. The signatures of the flying drone for these 2 different heights were quite similar due to the structure of the same drone. Figure 9 explains the time domain signal for the flying drone at a different height on the 2.6 GHz frequency at Taman Suria experiment site. The strength of signal for the 2.6 GHz frequency at Taman Suria was much lower compared to 1.8 GHz frequency. As shown in figure 9, the time taken for the drone to fly at the upper position was 1.6 seconds between the receiver and transmitter. It was completely different from the time taken by the drone to fly at the lower position, which only took 2.2 seconds.

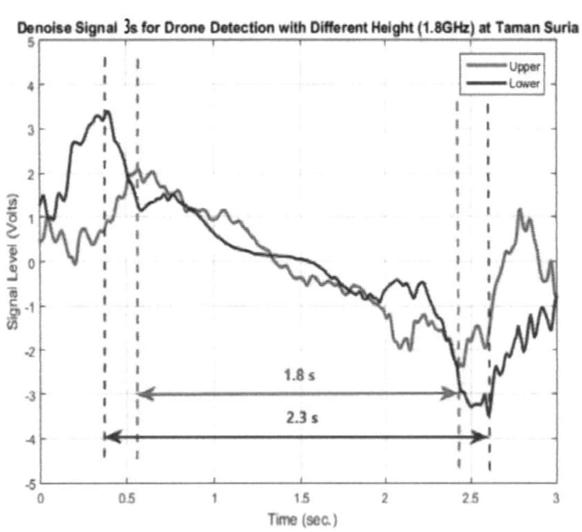

Figure 8. Time Domain of 1.8 GHz signal at Taman Suria.

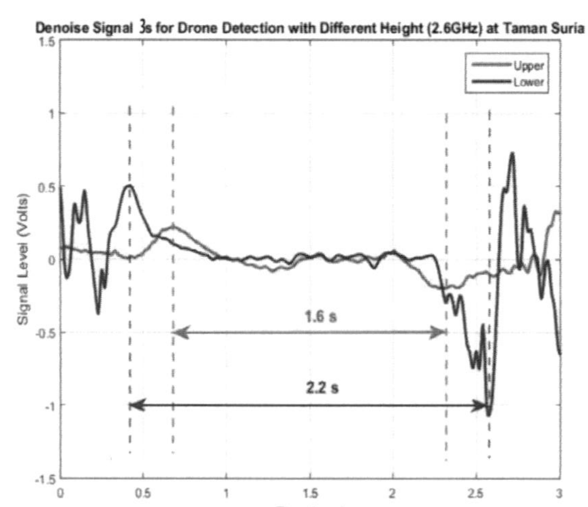

Figure 9. Time Domain of 2.6 GHz signal at Taman Suria.

In the meantime, figure 10 explains the time domain for detection on the 1.8G Hz frequency at Teluk Kemang. There were obvious differences in the time taken for the drone to fly at different heights. For the upper position, the drone took 1.9 seconds, while for lower positions, it only took 2.5 seconds. Unfortunately, at Teluk Kemang, there was only 1.8 GHz frequency of the signal that could be obtained from the transmitter. Although the signal level of 1.8 GHz at Teluk Kemang was quite lower than Taman Suria, it still can be used for detection experiment of a flying drone. This is because of the long distance between the transmitter and receiver. As we can see from figure 8, figure 9 and figure 10, the time taken for the drone to fly at the upper position was shorter compared to the lower position. This is because of the cross-sectional area of the FSR at the lower position being larger than the upper position.

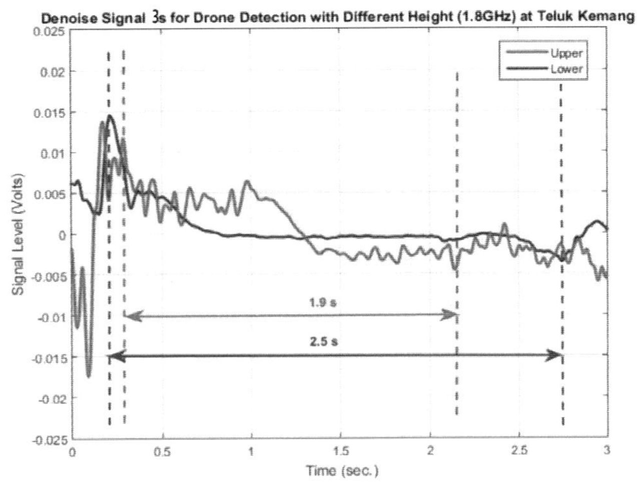

Figure 10. Time Domain of 1.8 GHz signal at Teluk Kemang.

3.2. Power Spectral Density of Drone and Prototype Detection

To analyze and measure the different frequencies for the signal, the Power Spectral Density (PSD) was used to change the graph with the signal. In definition, Power Spectral Density is the measure of signal power content versus frequency. For this study, PSD was specialized to characterize the signal of drone signature. The amplitude of the PSD is normalized by the spectral resolution employed to digitize the signal. In this analysis, it shows the analysis between Frequency (Hz) and Normalized Power (dB) of the flying drone at a different height. The results of PSD analysis is as shown in figure 11, figure 12 and figure 13. Based on the same drone used with the dimension 0.06755 m^2 for the upper and lower position of flying drone detection, the results affect the frequency dropped of the signature. In figure 11, the normalized power for the flying drone at the upper position was -40 dB and the frequency obtained was 1.1 Hz at Taman Suria with 1.8 GHz frequency of the signal transmitted. For the lower position, the normalized power was -90 dB and the frequency was 1.9 Hz. Figure 12 shows the PSD of 2.6 GHz signal for drone detection at Taman Suria. For the upper position, the normalized power was -50 dB and the frequency was 2.2 Hz. Lower position resulted in the normalized power of -28 dB and the frequency was 1.9 Hz.

For the last experiment at Teluk Kemang (1.8 GHz), the result of Power Spectral Density (PSD) is shown in figure 13. At the upper position of detection, the normalized power obtain was -38 dB and the frequency drop was 2.5 Hz. While for the lower position, the normalized power obtained was -60 dB and the frequency was 2 Hz. As we can see in Figure 11, figure 12 and figure 13, the frequency in the frequency domain graph was inversely related to the time in time domain graph. This is because of the relationship between frequency and time. When the frequency increases, the time will decrease.

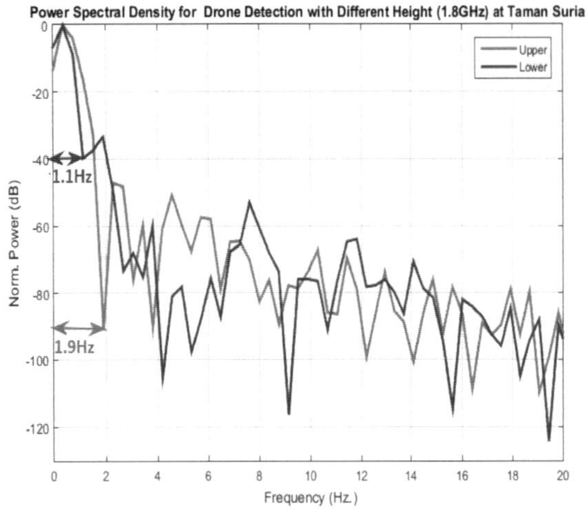

Figure 11. Power Spectral Density of 1.8 GHz signal at Taman Suria.

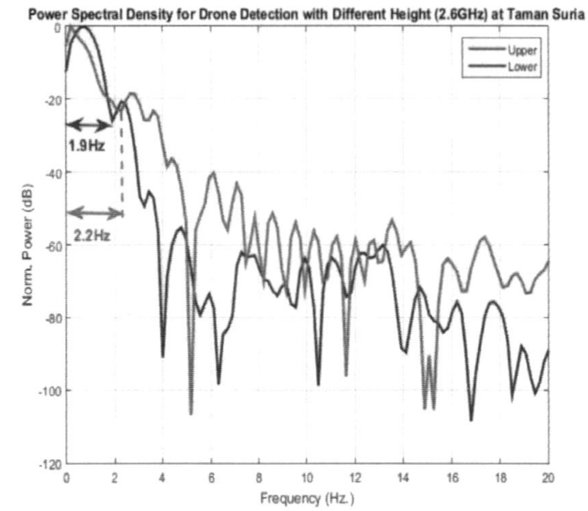

Figure 12. Power Spectral Density of 2.6 GHz signal at Taman Suria.

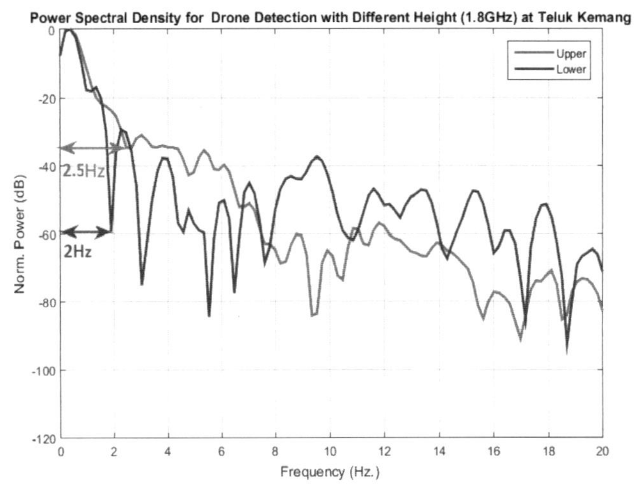

Figure 13. Power Spectral Density of 1.8 GHz signal at Teluk Kemang.

3.3. Principal Component Analysis

Principle Component Analysis (PCA) is used to convert some of the processed data into linearly uncorrelated variables, which are principle components. Through the PCA result, it can differentiate the pattern of the drone detection at different heights at the specific area. The result of PCA at Taman Suria for 1.8 GHz and 2.6 GHz are shown in figure 14 and figure 15. The blue dots indicate the lower position, and the upper position is indicated by the red dots. Based on figure 14, the cluster perfectly classified the data, and figure 15, the result of the PCA shows that the clusters for two different heights overlap each other. The signal strength and quality for frequency signal of 1.8 GHz were stronger and better quality with low noise compared to 2.6 GHz. Moreover, the good quality signal at Taman Suria was because the transmitter was located near the radar receiver at only 187.5 meters away. In addition, the good blockage by the building around prevented the signal from being interrupted by another signal from other transmitter towers, which lead to good quality signal received.

Next, the result of PCA at Teluk Kemang is shown in figure 16 for the signal of 1.8 GHz only because there was no signal of2.6 GHz present near the experiment site. The analysis can be classified but was not as good as at Taman Suria. The clusters almost overlap due to the dimension that was almost the same, but the signal quality at Teluk Kemang was not strong and good enough to get better result. One of the reasons was that the distance between the transmitter and receiver was too far away at 580 meters. Moreover, the area of the experiment site was an open space and had no blockage by the surrounding buildings, which can lead to the signal being interrupted by other transmitter tower signal that can lead to interference during the experiment.

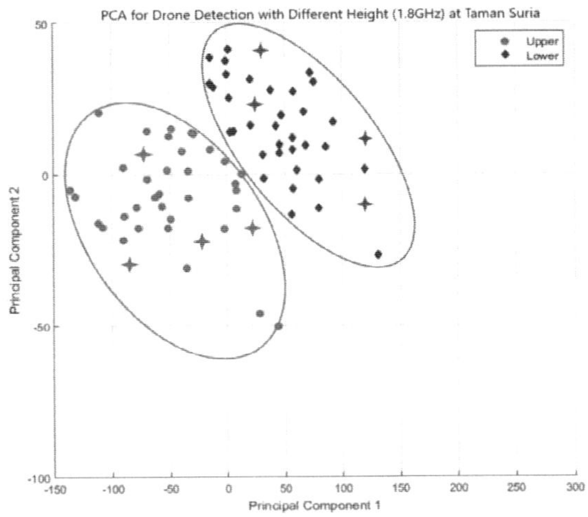

Figure 14. Principal Component Analysis of 1.8 GHz signal at Taman Suria.

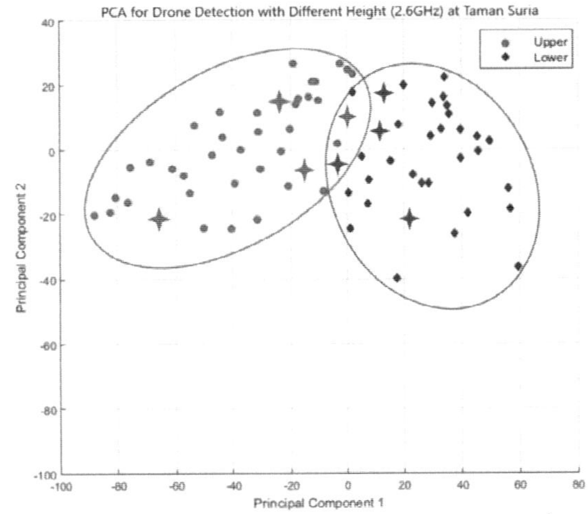

Figure 15. Principal Component Analysis of 2.6 GHz signal at Taman Suria.

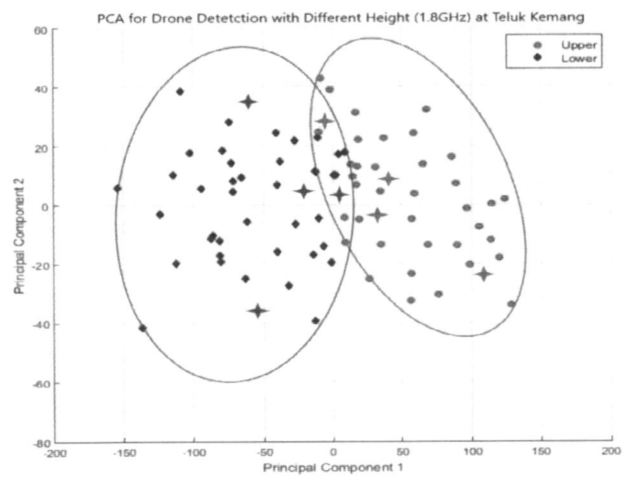

Figure 16. Principal Component Analysis of 1.8 GHz signal at Teluk Kemang.

3.4. Confusion Matrix

To determine the performance and precision of the classifier, the confusion method was applied in this analysis. All of the analysis was tabulated and in the form of percentage. Table 3, table 4 and table 5 are the results of the confusion matrix for the drone detection at different heights at different frequencies and sites. Table 3 tabulates the result for the confusion matrix with a frequency of 1.8 GHz at Taman Suria. From the PCA result, all of the testing data managed to stay in its cluster. It was resulting 100% success testing data. For the frequency of 2.6 GHz at Taman Suria, the result of the confusion matrix is shown in table 4. 75% of the tested data for the upper and lower detection successfully stayed in their own clusters. It has been proven from the result of PCA. Only one data did not stay in its cluster, which is 25% of the tested data. Data collection at Taman Suria for both frequencies was quite good and satisfactory. This was due to the strong transmission of signal from the transmitter tower and there were less obstacles.

For the Teluk Kemang confusion matrix result, it is tabulated in table 5. There were 4 data tested for each upper drone detection and lower drone detection. The result shows only one data that did not stay in its cluster while the other three data stayed in their clusters. In percentage, 75% of the tested data successfully stayed in their clusters, while the other 25% did not.

Table 3. Confusion matrix for 1.8 GHz at Taman Suria.

Level of height	Number of samples		Automatically classified (100%)	
	Training	Testing	Upper	Lower
Upper	20	4	**100**	0
Lower	18	4	0	**100**

Table 4. Confusion matrix for 2.6 GHz at Taman Suria.

Level of height	Number of samples		Automatically classified (100%)	
	Training	Testing	Upper	Lower
Upper	21	4	**75**	25
Lower	19	4	25	**75**

Table 5. Confusion matrix for 1.8 GHz at Teluk Kemang.

Level of height	Number of samples		Automatically classified (100%)	
	Training	Testing	Upper	Lower
Upper	18	4	**75**	25
Lower	19	4	25	**75**

4. Conclusion

In conclusion, it shows clearly the successful results of classification of data between the detection of flying drone at upper and lower position in Principle Component Analysis (PCA). By comparing the two sites of experiment, using the LTE signal and passive forward scatter radar hardware system, detection of drone at Taman Suria at a frequency of 1.8 GHz was most successful. This was proven in the result of confusion matrix of testing data, where all of the testing data managed to stay in its cluster. Furthermore, the distance between the transmitter antenna and receiver antenna also affected the results of data analysis. The higher the distance between transmitter and receiver, the lower the quality of signal received.

The great potential of this passive forward scatter radar system provides a new research area in passive radar that can be used for diverse remote monitoring, including border protection, microwave fences, building monitoring, traffic surveillance, and so forth. Henceforward, future LTE based passive forward scattering radar investigations may consider air targets such as helicopters, and sea targets such as battle ships, humans and animals.

For the future recommendation, the passive forward scattering radar system could use the signal from transmitter with 5 Generation Wireless System (5G) network that provide stronger, better quality and wider network coverage signal which can detect drone precisely and give the result of Doppler signature more clearly.

References

[1] Kadhim A S 2017 Design ground based radar by using MALTLAB Simulink *International Journal OF Engineering Sciences &Management Research* **5** no 4 p 38

[2] Raja Abdullah R S A and Ismail 2006 A forward scattering radar: current and future applications *Int. J. Eng. Technology* **3** no 1 p 61-7

[3] Diana A 2018 Pecah rumah guna dron *Utusan Malaysia* 15 Nov

[4] Abdul Aziz N H, Raja Abdullah R S A and Mohd Yusof A N 2017 Human detection and recognition system using passive forward scattering radar *Science International* **29** no 1 p 69–73

[5] Raja Abdullah R S A, Salah A A, Ismail A, Hashim F and Abdul Aziz N H 2016 Experimental investigation on target detection and tracking in passive radar using Long-Term Evolution signal *IET Radar, Sonar & Navigation* **10** issue 3 p 577-85

[6] Raja Abdullah R S A, Abdul Aziz N H, Salah A A, Hashim F 2016 Analysis on target detection and classification in LTE based passive forward scattering radar *Sensors* **16** no 10 p 1607

[7] Jolliffe I T and Cadima J 2016 Principal component analysis: a review and recent developments *Philosophical Transactions of the Royal Society A: Mathematical, Physical and Engineering Sciences* 374 no 2065 p 20150202

Journal of Physics: Conference Series 2250 (2022) 012004 doi:10.1088/1742-6596/2250/1/012004

Slot implementation on second iteration of fractal log periodic dipole antenna for UHF band application

Muhammad Hafeez Rosmin, Mohamad Kamal A Rahim, Nur Syahirah Yaziz, Muhammad Naeem Iqbal, Osman Ayop, Muhammad Amir Haziq Kamarusshahrin and Farid Zubir

Advanced RF and Microwave Research Group (ARFMRG), Communication Engineering Department, Faculty of Electrical Engineering, Universiti Teknologi Malaysia, 81310 UTM Johor Bahru, Johor, Malaysia

E-mail: mdkamal@utm.my

Abstract. In this paper, the design of the second iteration of Log Periodic Fractal Koch Antennas (LPFKA) with and without slot implementation is designed for Ultra High Frequency (UHF) band applications. The slot implementation is applied on each of the radiating elements to reduce the antenna's size and to prevent the shifting of a lower frequency to a higher frequency. The antenna is operated at a frequency between 0.5 GHz and 9.0 GHz. The Computer Simulation Technology (CST) software was used to design and simulate the antenna. The performance of the antenna such as reflection coefficient, radiation pattern and gain have been analyzed. The antennas have been fabricated using an FR4 laminated board with a dielectric constant of 4.6, tangent loss of 0.019, and a thickness of 1.6mm. The results show a positive outcome, with a stable radiation pattern over the operational bandwidth and a reflection coefficient of less than -10 dB for the designed frequency.

1. Introduction
Wireless communications are now one of the trendiest topics in rapid development because it offers enhanced network interaction, data transmission, video call and other services. This has led to various kind of antenna research.

One of them employs fractal-shaped antenna elements [1]. Most of these designs have shown to be especially effective in terms of lowering antenna size, while others try to incorporate multi-band properties [2–5]. Antenna elements with a fractal shape bring a number of advantages such as antenna size reduction, broad bandwidth and multiband frequency [6]. An example of a good self-similar fractals shape that has been used to construct multi-band and miniaturized antennas is Koch curve. Most of the constraints of small antennas can be solved by the Koch geometry's properties [7].

For this paper, we emphasis on the size of the antenna therefore a Koch curve shape is applied. For Koch curve shape, there are zero iteration which is the conventional shape of log periodic dipole antenna, first iteration, series iteration and second iteration. The more iteration it has, the more size reduction can be obtained. Even so, as the number of iterations increases, the operating frequency will shift more to the higher frequency. In order to avoid that, the slot implementation is done [8]. In this paper, we only analyze the slot implementation on the second iteration.

In theory, the slot implementation was done to improve the bandwidth of the antenna, for size reduction as well as to prevent the shifting of the lower frequency to the higher frequency.

This paper discusses the antenna design process by applying the log periodic technique, as well as the detailed results such as reflection coefficient, radiation pattern, and gain of the antenna design.

2. Antenna Design

Log periodic principles are used to construct fractal Koch antennas for the 2nd iteration. The scaling factor (τ), which is described as the ratio between 2 consecutive antenna elements in terms of length (l), width (w) of antenna elements, number of the elements (n) and distance between antenna elements (d) are given in Equation (1), is among the key parameters in constructing the log periodic antennas [9].

$$r = {d_{n+1}}/{d_n} = {l_{n+1}}/{l_n} = {w_{n+1}}/{w_n} \quad (1)$$

Initially, both antennas are constructed and modelled by using Computer Simulation Technology software as shown in figure 1. Then, the antenna is fabricated by using FR-4 laminated board with dielectric constant of 4.6, tangent loss of 0.019 and thickness of 1.6mm as shown in figure 2.

Each side of the antenna has 12 elements that have been lined in a crossed arrangement in order to generate enough electromagnetic radiation. The value of the scaling factor is 0.85 by referring to Carrel's table [9]. One side of the antenna is connected to the excitation port and another connected to the ground port. Both antennas are done at a 60-degree angle of second iteration.

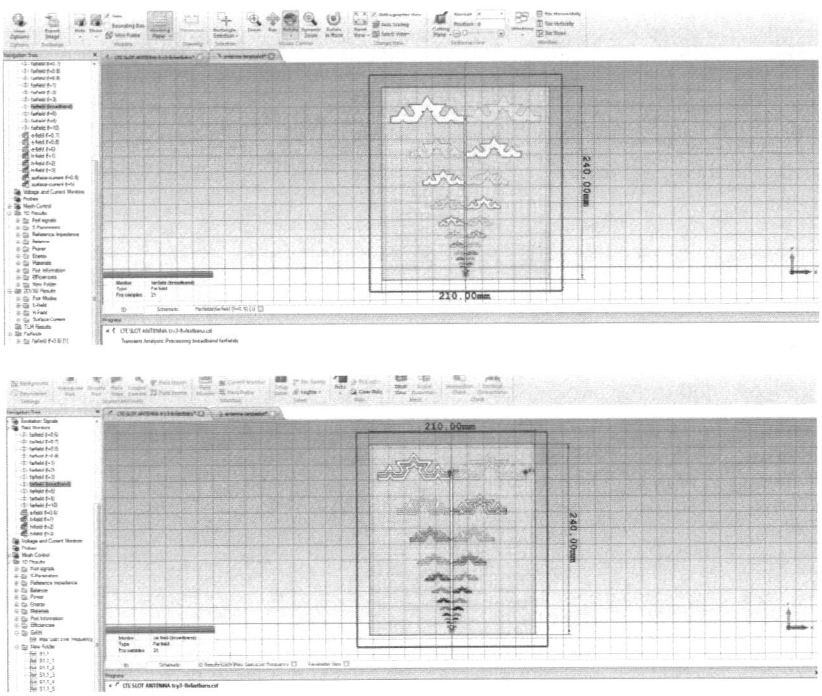

Figure 1. Design of standard and slotted 2nd iteration antenna in CST software.

Journal of Physics: Conference Series 2250 (2022) 012004 doi:10.1088/1742-6596/2250/1/012004

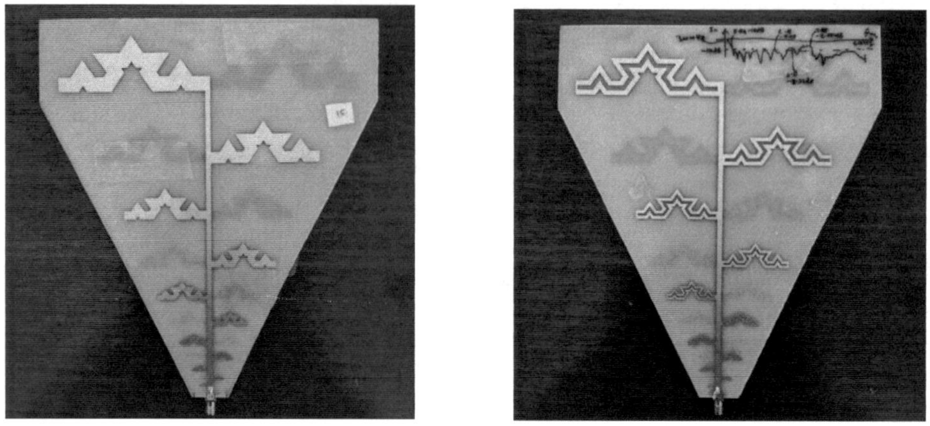

a) Standard 2nd iteration b) Slotted 2nd iteration

Figure 2. Show the fabricated antennas.

3. Results and analysis

This section presents the outcomes for both antennas with respect to their reflection coefficient, radiation pattern and gain.

3.1. Reflection Coefficient

Figure 3 show the simulated and measured reflection coefficient for the standard 2nd iteration and slotted 2nd iteration antennas. From the simulated readings, it can be seen that even though we implement the slot, the reflection coefficient of both antenna doesn't really change much and in fact much similar to each other. Figure 4 show the comparison of simulated and measured reflection coefficient for each antenna. From both figures, the antennas demonstrate a satisfactory response with respect to -10 dB over the desired frequency.

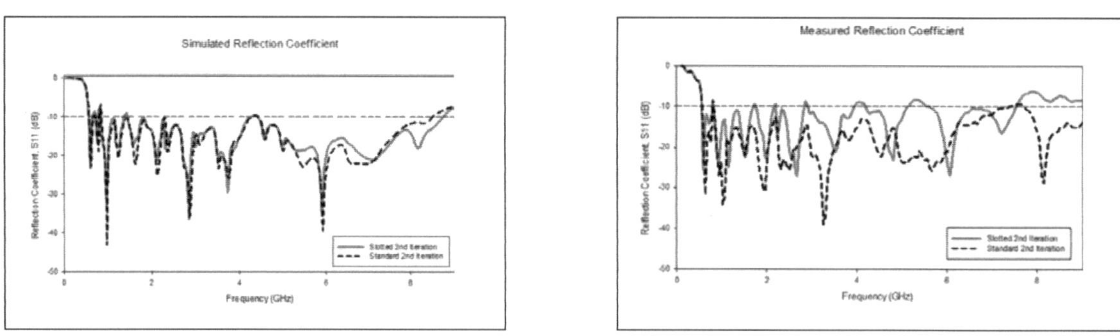

a) Simulated reflection Coefficient b) Measured reflection coefficient

Figure 3. Simulated and measured reflection coefficient for standard 2nd iteration and slotted 2nd iteration of fractal Koch antenna.

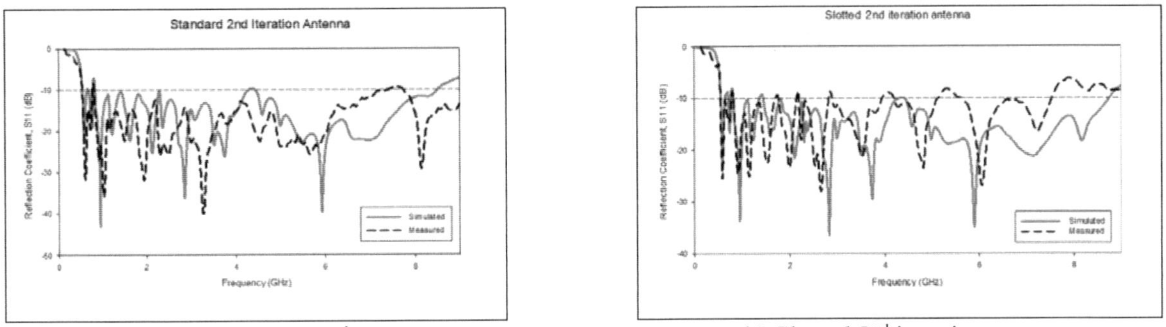

a) Standard 2nd iteration b) Slotted 2nd iteration

Figure 4. Comparison of simulated and measured reflection coefficient for each antenna.

From the reflection coefficient (S11) of both antennas, the bandwidth improvement is observed. From Table 1, it can be seen that the slot implementation improves the bandwidth performance as well as prevent the frequency shifting due to second iteration implementation.

Table 1. Bandwidth improvement.

Type of antenna (FKLPA)	Operating Frequency (GHz)	Bandwidth Improvement (MHz)
Standard 2nd iteration	0.57 – 8.50	-
Slotted 2nd iteration	0.55 – 8.80	320

3.2. Radiation Pattern

This section discusses the radiation patterns for both antennas. Figure 5 shows the simulated 3D radiation pattern for standard 2nd iteration while figure 6 for slotted 2nd Iteration. Figure 7 and figure 8 shows the simulated and measured 2D radiation pattern for standard 2nd iteration antenna and slotted 2nd iteration antenna respectively. Both antennas are evaluated at 0.9 GHz, 1 GHz, 2 GHz and 3 GHz. From both the 2D and 3D radiation patterns, it can be seen that all antennas radiate towards the smaller elements resulting in directional patterns. All of the 2D and 3D radiation patterns shows that the simulated and measured readings is almost similar.

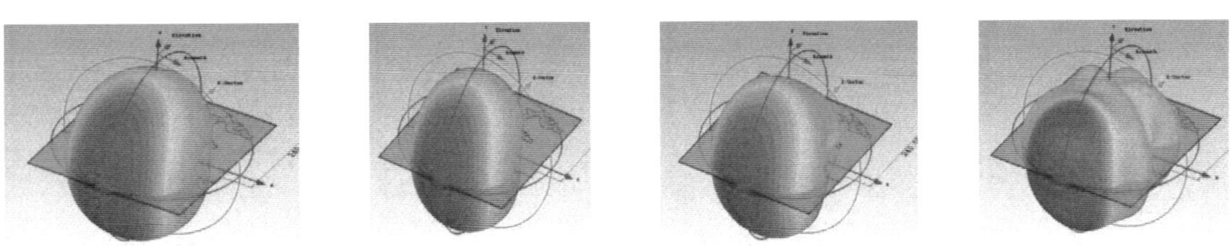

a) Frequency at 0.9 GHz b) Frequency at 1 GHz c) Frequency at 2 GHz d) Frequency at 3 GHz

Figure 5. Simulated 3D radiation pattern for standard 2nd iteration antenna

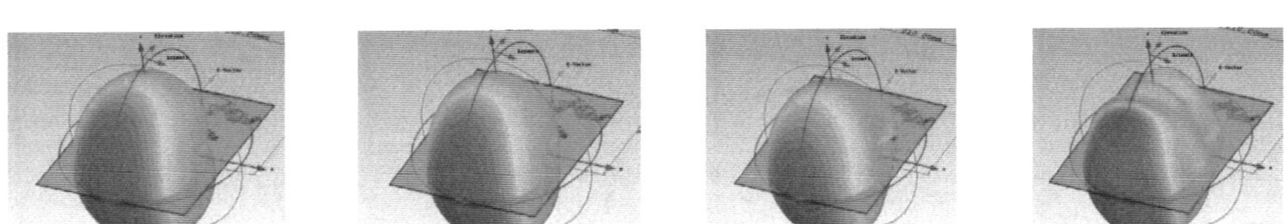

a) Frequency at 0.9 GHz b) Frequency at 1 GHz c) Frequency at 2 GHz d) Frequency at 3 GHz

Figure 6. Simulated 3D radiation pattern for slotted 2nd iteration antenna

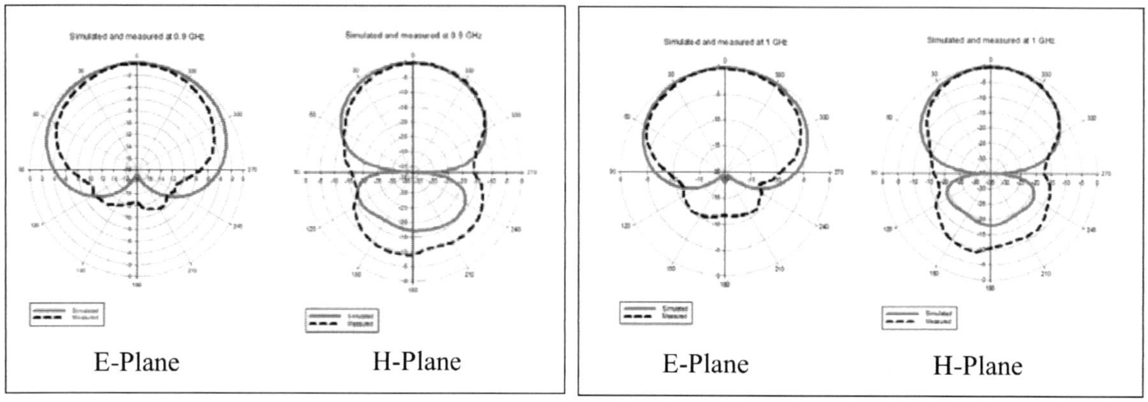

a) Frequency 0.9 GHz b) Frequency 1 GHz

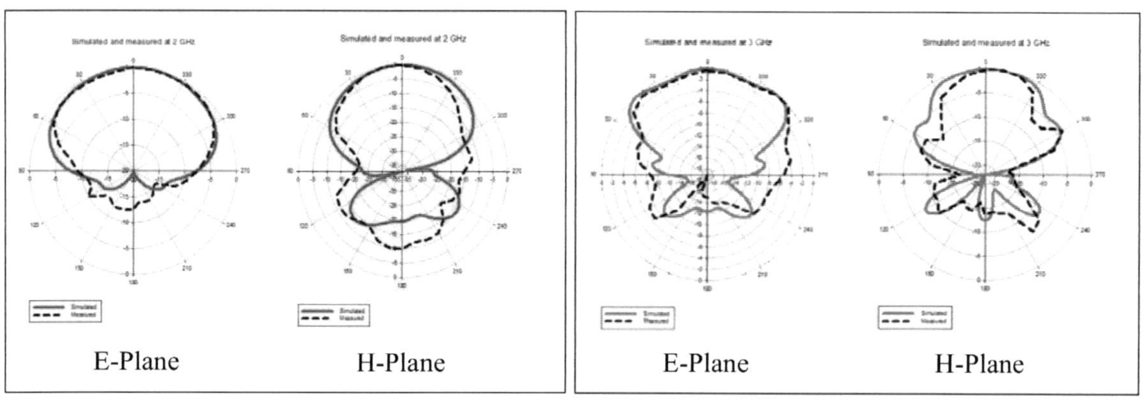

c) Frequency 2 GHz d) Frequency 3 GHz

Figure 7. Simulated and measured 2D radiation pattern for standard 2nd iteration Fractal Koch antenna.

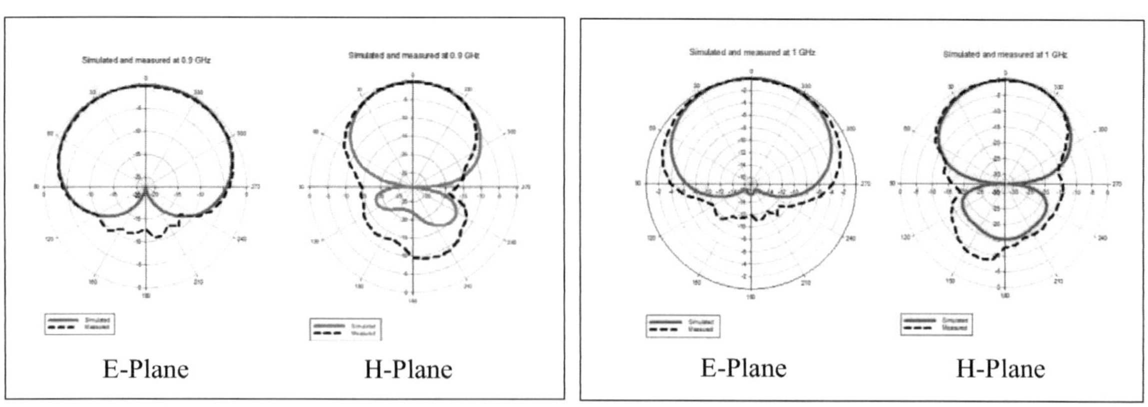

a) Frequency 0.9 GHz b) Frequency 1 GHz

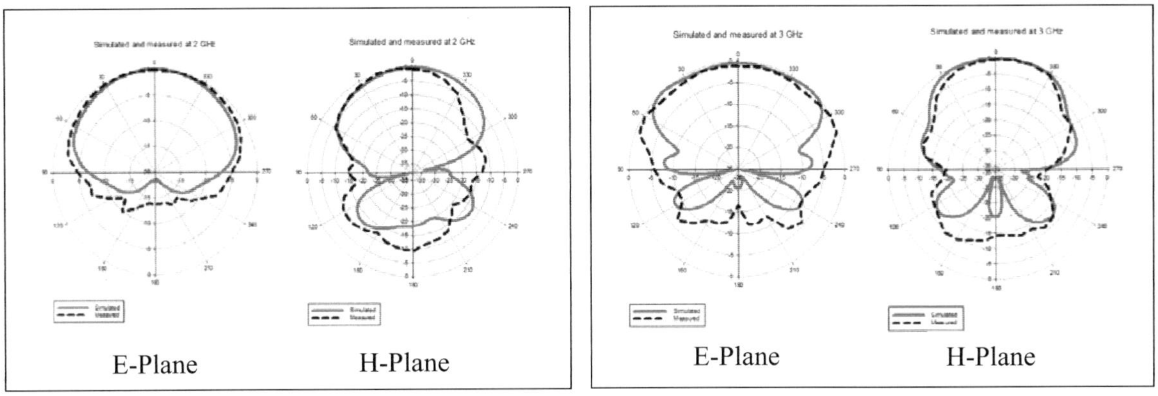

c) Frequency 2 GHz d) Frequency 3 GHz

Figure 8. Simulated and measured 2D radiation pattern for slotted 2nd iteration Fractal Koch antenna.

3.3 gain

Figure 9 shows the simulated realized gain for standard 2nd iteration antenna and slotted 2nd iteration antenna. Both antennas give good result that maintain higher than 5dBi throughout the designed frequency which is from 0.5GHz until 9GHz. For directional antenna, the gain must be higher than 5dBi to consider it a good antenna in order to reduce the energy loss from the antenna.

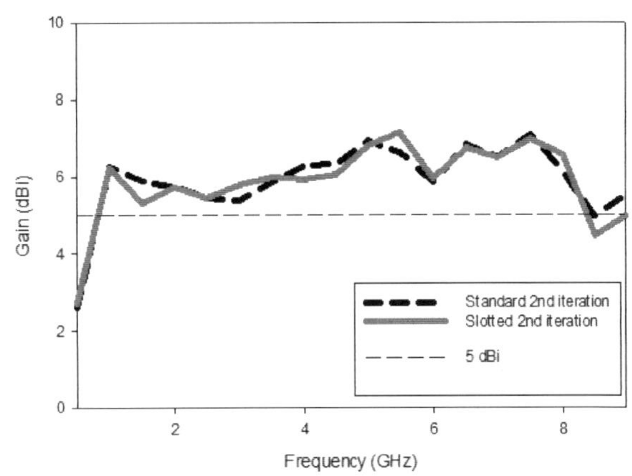

Figure 9. Simulated realized gain for both antennas.

4. Conclusion

In this paper, 2nd iteration fractal log periodic dipole antenna with and without slot implementation has been proposed. Both antennas show a good reflection coefficient result with respect to -10 dB for the frequency between 0.5 GHz and 9.0 GHz. As the number of iterations increases, the shifting of lower frequency has been restored by implementing slot to each radiating elements of the antenna. The slotted 2nd iteration antenna has 320MHz bandwidth improvement compared to the standard 2nd iteration antenna. From the result, the proposed antenna can be used for many applications.

Reference

[1] Vinoy, K. J. *Fractal shape antenna elements for wide band and multi band wireless applications* Ph.D. Thesis, The Pennsylvania State University, August 2002

[2] Cohen, N. *Fractal antenna applications in wireless telecommunications* Professional Program Proc. of Electronics Industries Forum of New England, 1997, IEEE, 43–49, 1997.

[3] Puente-Baliarda, C., J. Romeu, R. Pous, J. Ramis, and A. Hijazo, *Small but long Koch fractal monopole* Electron. Lett., Vol. 34, 9–10, 1998.

[4] Kim, L. K., J. G. Yook, and H. K. Park, *Fractal shape small size microstrip patch antenna* Microwave and Optical Technology Letters, Vol. 34, No. 1, Jul. 5, 2002.

[5] Gianvittorio, J. P. and Y. Rahmat-Samii, *Fractal antenna: A novel antenna miniaturization technique, and applications* IEEE Antenna's and Propagation Magazine, Department of Electrical Engineering, University of California, Los Angeles Engr. IV, Los

[6] L.K. Kim, J.G. Yook, and H.K. Park *Fractal shape small size microstrip patch antenna* Microwave Opt Technol Lett 34 (2002).

[7] Ramadan, A., K. Y. Kabalan, A. El-Hajj, S. Khoury, and M. AlHusseini *A reconfigurable U-koch microstrip antenna for wireless applications* Progress In Electromagnetics Research, PIER 93, 355–367, 2009.

[8] R. Dakir, J. Zbitou, A. Mouhsen, A. Tribak, M. Latrach, and A. M. Sanchez, *A New Compact and Miniaturized GCPW-fed Slotted Rectangular antenna for Wideband UHF FIRD Applications* Int. J. Electr. Comput. Eng., vol. 7, p. 767, 2017

[9] C. A. Balanis, *Antenna Theory Analysis and Design*, Third Edition, John Wiley and Sons, New York, 2005

[10] N.S.M. Yaziz , M.K.A Rahim , F. Zubir , N.S. Nadzir , R. Dewan, H.A. Majid. *The Improvement of first Iteration Log Periodic Fractal Koch Antenna with Slot Implementation* 2018

[11] Inkwinder Singh Bangi1 · Jagtar Singh Sivia1. *Moore, Minkowski and Koch Curves Based Hybrid Fractal Antenna for Multiband Applications* 2019.

[12] Dmitrii Tumakov, Dmitry Chikrin and Petr Kokunin. *Miniaturization of a Koch-Type Fractal Antenna for Wi-Fi Applications* 2020.

[13] G. Varamini, A. Keshtkar, N. Daryasafar, and M. Naser -Moghadasi, *Microstrip Sierpinski fractal carpet for slot antenna with metamaterial loads for dual-band wireless application* AEU - Int. J. Electron. Commun., vol. 84, pp. 93–99, 2018.

[14] Yichen Xiong, Bing Zhang, *A miniaturized three-dimensional log periodic Koch-dipole array antenna using T-shaped top loading* International journal of RF and Microwave computer aided engineering, Vol. 31, Issue 7, 2021.

Acknowledgement
The authors would like to thank the Ministry of Higher Education (MOHE), School of Postgraduate Studies (SPS), Research Management Centre, Advanced RF and Microwave Research Group, School of Electrical Engineering and Universiti Teknologi Malaysia (UTM), Johor Bahru, for the support of the research under Grant 4J449, 4B588 and 06G15.

Journal of Physics: Conference Series 2250 (2022) 012005 doi:10.1088/1742-6596/2250/1/012005

Biocompatible Sago Aerogel Material as A Substrate for Antenna Application

M A Adnan[1], K Kamilia[1,2], Y Yamada[1], I H Idris[3], A A Ali[4] and H. Hamdan[4]

[1] Communication Systems and Networks Research Lab, Malaysian-Japan International Institute of Technology (MJIIT), Universiti Teknologi Malaysia, Kuala Lumpur, Malaysia

[2] Wireless Communication Centre, Universiti Teknologi Malaysia, Kuala Lumpur, Malaysia

[3] Advanced RF & Microwave Research Group (ARFMRG), School of Electrical Engineering (SKE), Faculty of Engineering (FE), Universiti Teknologi Malaysia, Johor Bahru, Johor, Malaysia

[4] Razak Faculty of Technology and Informatics, Universiti Teknologi Malaysia, Kuala Lumpur, Malaysia

Corresponding authors' email: kamilia@utm.my

Abstract. This paper presents a patch antenna that operates at 2.4 GHz for body communication applications. New biocompatible material is proposed named Sago Aerogel is used as the antenna substrate. As a new substrate material is proposed, a simple patch antenna with a total size of substrate size of 60 mm × 66.2 mm × 1.6 mm size designed and simulated. The antenna performances in terms of S_{11} value, radiation pattern, gain, and input impedance are studied and discussed.

1. Introduction

The use of antennas in this century has grown considerably in both the telecommunications and medical sectors. In the medical sector with the use of an antenna, continuous monitoring of patients can be made and fast transmission of data is possible and more sophisticatedly is by using an antenna in biomedical implants for detecting various illnesses in one's body at every moment. With the introduction of Implantable Medical Devices (IMD) and biomedical implants, the growth of the world population has the potential to be increased and at the same time allow the increase of global aging as well. In IMD, the antenna that is suitable to be implanted must be small in size and biocompatible in characteristics.

To apply an antenna in a biomedical application system, biocompatible material of antenna substrate needs to be used. It is to ensure that the antenna did not produce any harm to the human body. Biocompatible materials are synthetic or natural materials that are used to replace parts of the living system or function in intimate contact with living tissue. Biocompatible materials are intended to connect with the biological system to evaluate, treat, supplement or replace any tissue, organ, or function. Biocompatible materials differ from biological materials such as bone produced by biological systems. Artificial coatings, vascular storage, artificial appliances, and catheters are all made from different biological materials and are made up of different medical devices [1].

Previous work on biocompatible material is reported in [2-9]. The performance antenna in [2] is compared between three different substrates which are Rogers RO 3210, Silastic 0.2 mm, and Alumina. The antennas are intended to be used in intracranial pressure (ICP) monitoring applications. As a result, the Alumina substrate shows the best performance compared to the others substrate. Another reported work using Alumina substrate is presented in [3]. Rogers 3210 also has been used in [4] while using Polydimethylsiloxane (PDMS) as the superstrate. Antenna in [5] used Taconic as a substrate and PDMS as the superstrate. Other reported substrates that have been used for the biocompatible antenna are Kapton polyimide [6], TiN film [7], and ULTRALAM [8-9].

However, new material is proposed in this paper which is Sago Aerogel. Sago is a starch source that can be used for bioplastic synthesis. Sago starch is derived from Metroxylon Sago palm oil, which is a commonly found palm tree exported by countries in Southeast Asia [10]. Aerogel is a synthetic porous ultralight material derived from gels, in which the liquid components of the gel have been replaced with gas. The result is compact with very low density and very low heat conductivity. Other names include frozen smoke, solid smoke, solid air, solid clouds, blue smoke due to its translucent properties and the way it absorbs light in materials [4].

This new material is proposed as a new material since it is easy to be fabricated. The permittivity and tangent loss of the Sago substrate is measured. Then, the patch antenna is simulated using Sago substrate to obtain the antenna performance. The antenna performance using Sago substrate will be compared using the common substrate material which FR-4 and Teflon as the references.

2. Antenna Design and Substrate Details

The material that is being used as substrate can affect the patch antenna's gain, return loss, directivity, resonance frequency, bandwidth as well as the dimension of the patch of the antenna (length of patch, width of patch, and inset depth) [11]. Figure 1 and figure 2 show the proposed substrate that is being used in this project which is by using sago aerogel. The permittivity and loss tangent of the substrate is measured using Vector Network Analyzer (VNA) and the result obtained is shown in table 1.

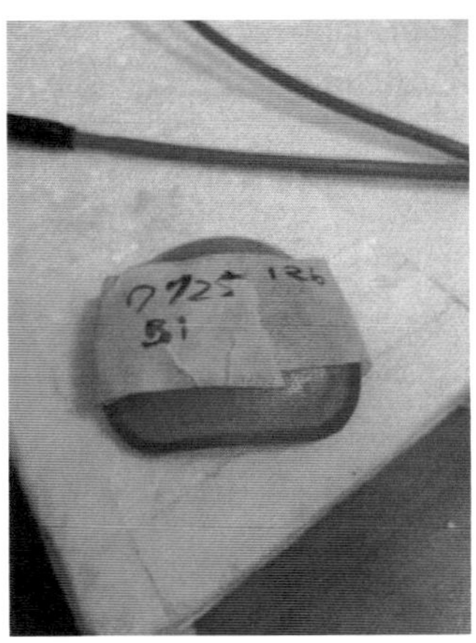

Figure 1. Sago aerogel material (front view).

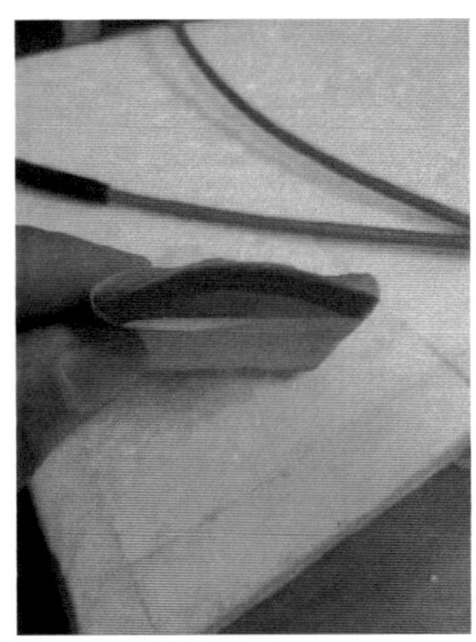

Figure 2. Sago aerogel material (side view).

The patch antenna is designed as shown in figure 3, using theoretical equations as in reference [3]. By using the value of permittivity, $\varepsilon = 2.72$, and tangent loss, $\tan\delta = 0.057$ as stated in table 1, the calculated parameters for the patch antenna are obtained. The results of FR-4 and Teflon as the reference substrates are needed to compare with Sago. The patch antenna model is shown in figure 3. The calculated parameters for FR-4, Teflon, and Sago are shown in table 1.

3. Results and Discussion

Figure 4 shows the S_{11} results for three different substrates used in this paper. First is S_{11} result with FR-4 substrate is shown in figure 4. As shown in the figure, the antenna can operate at 2.4 GHz (-22.21 dB) with the 70 MHz bandwidth value (2.44 GHz to 2.37 GHz). Then, the S_{11} result using Teflon substrate. The antenna can operate at 2.4 GHz (-15.11 dB) with the 40 MHz bandwidth value (2.42 GHz to 2.38

GHz). Lastly, the S_{11} result using Sago substrate. The proposed antenna can operate in 2.4 GHz (-19.14 dB) with the 40 MHz bandwidth value (2.42 GHz to 2.38 GHz). Comparison of S_{11} results by using three different substrate materials shows that the Teflon and Sago substrates have similar bandwidth values but the Sago substrate has a better S_{11} value compared to the Teflon substrate. By observing the figure, the higher the permittivity of the substrate gives lower return loss for the antenna.

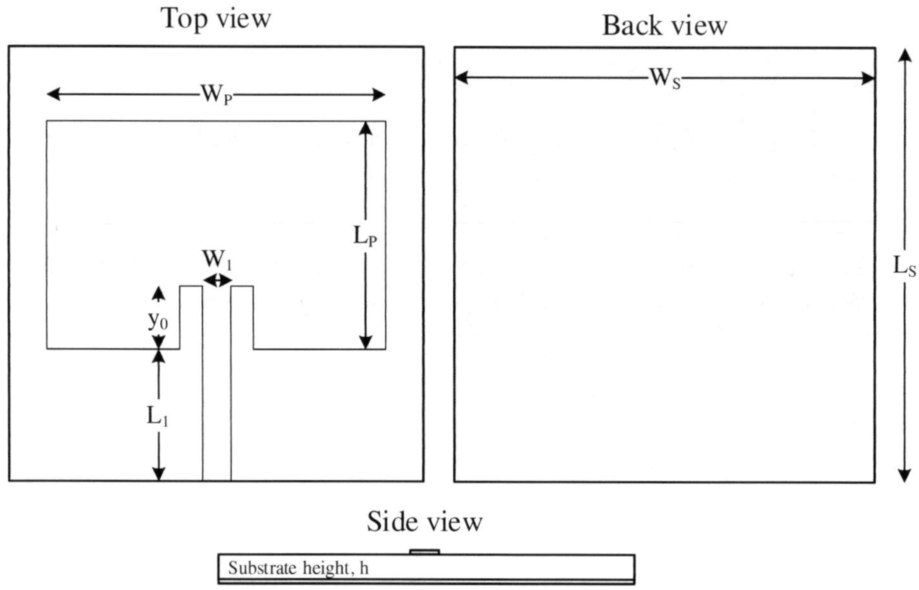

Figure 3. Antenna geometry of the Microstrip Patch Antenna.

Table 1. Design parameters using different types of substrates.

Parameter	FR-4	Teflon	Sago
Dielectric Constant, ε	4.3	2.1	2.72
Loss Tangent, tan (δ)	0.01	0.00028	0.057
Substrate Length, L_S	60 mm	87 mm	66.181 mm
Substrate Width, W_S	50 mm	70 mm	60 mm
Substrate Thickness, h	1.6 mm	1.6 mm	1.6 mm
Patch Length, L_P	29 mm	41 mm	30.18 mm
Patch Width, W_P	28 mm	50 mm	38.46 mm
Inset Length, y_0	9 mm	10 mm	9.365 mm
Microstrip Line Length, L_l	15 mm	22 mm	17 mm
Microstrip Line Width, W_l	2 mm	5.5 mm	3.09 mm

Next, figure 5 (a) shows the radiation pattern results for all three different substrates at 2.4 GHz used in this paper. As seen in the figure, the patterns for all substrates are pointing in an upward direction. A small back lobe is noticed in the pattern results. 3D pattern for Sago substrate is shown in figure 5 (b). As noticed, most of the radiation intensity is pointing upward of the patch antenna whereas the lower radiation intensity is towards the ground plane.

Simulated gain results at 2.4 GHz are tabulated in table 2. The highest total gain of the patch antenna is Teflon with 7.23 dBi. Sago and FR-4 as substrate can also be considered as a good antenna performance because both of the substrates reach the optimum gain.

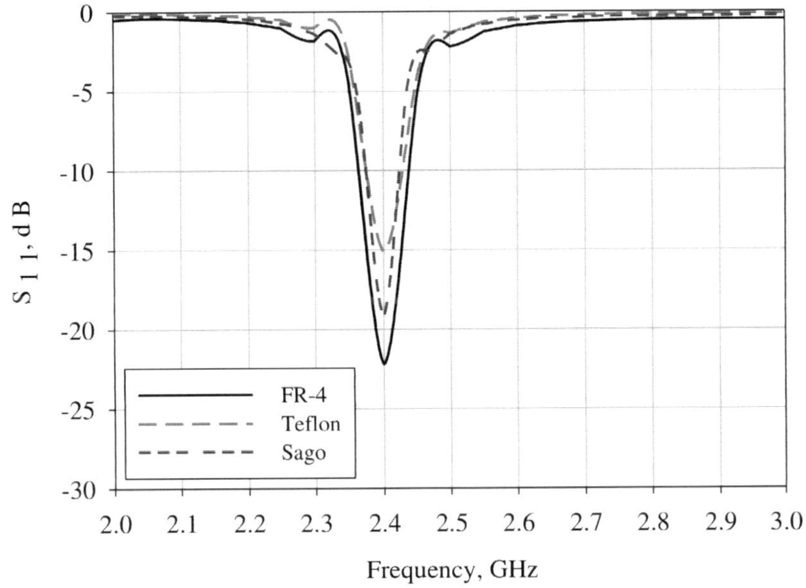

Figure 4. Comparison of simulated S_{11} results between all substrates.

Table 2. Simulated gain results of patch antenna using different types of substrates.

Substrate Materials	Simulated Gain, dBi
FR-4	3.75
Teflon	7.27
Sago	4.97

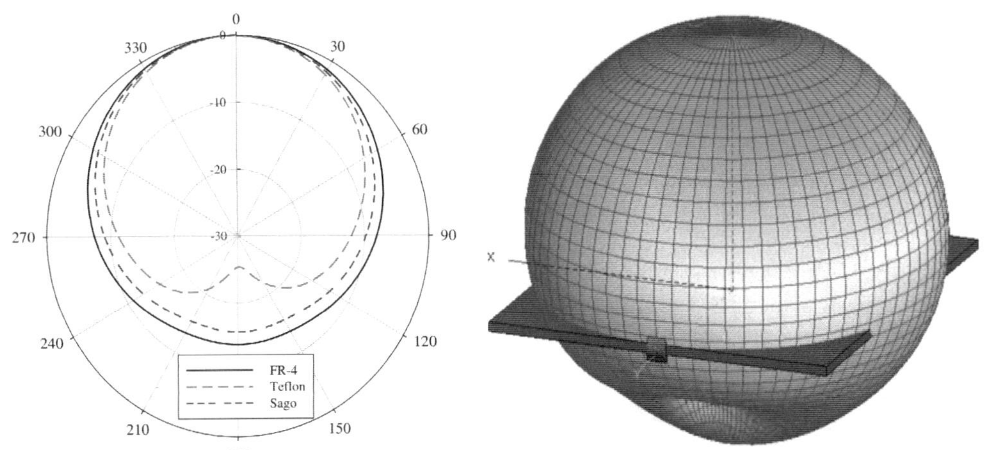

(a) (b)

Figure 5. (a) Simulated radiation pattern results for all substrates (b) 3D radiation pattern for Sago substrate.

Figure 6 shows the input impedance results for three different substrates used in this paper. The ideal input impedance for the patch antenna is 50 Ω with 0 imaginary value. Input impedance at 2.4 GHz using FR-4, Teflon, and Sago substrates is 44.51+4.87j, 40.17+12.61j, and 45.83+9.78j, respectively. Although it is impossible to get the ideal input impedance, getting the closest value to the ideal can be considered as a good antenna. From the figures, all substrates attained a closed value to the ideal input impedance.

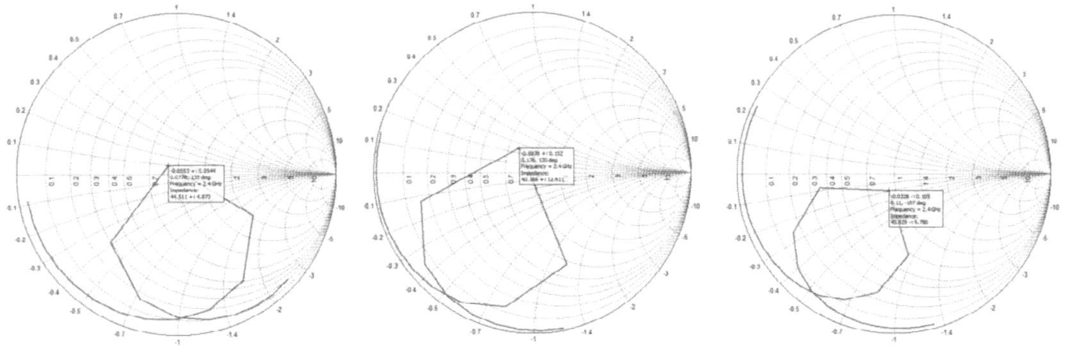

(a) (b) (c)

Figure 6. Simulated input impedance using (a) FR-4 (b) Teflon (c) Sago substrates

4. Conclusion

The results obtained from the research shows that Sago as a substrate is suitable to be used as a biocompatible antenna. The results of the Sago substrate compared to commonly used substrates such as FR-4 and Teflon are on the same level of performance. Good antenna performance results when using Sago substrate is obtained in terms of S_{11} value, radiation pattern, gain and input impedance. However, a detailed simulation study needs to be carried out since it is a challenge to have a flat surface of Sago substrate in real environment. Fabricating a patch antenna using Sago substrate is required to test the antenna in a real environment.

5. References

[1] Ng I C, Pawijit P, Tan J and Yu H 2019 Anatomy and Physiology for Biomaterials Research and Development *Encyclopedia of Biomedical Engineering* p 225-236

[2] Psathas K A, Kiourti A and Nikita K S 2014 Biocompatibility of implantable antennas: Design and performance considerations *The 8th European Conference on Antennas and Propagation (EuCAP)* p 1566-1570

[3] Mokhtar M, Kamardin K, Yamada Y, Baharin R H M, Rahman N H A, Zainudin N, Latef T A and Yasin M N M 2018 Implantable Patch Antenna for Body Communication *2018 2nd International Conference on Telematics and Future Generation Networks (TAFGEN)* p 77-80

[4] Loktongbam P and Laishram R 2017 Performance and Design Analysis of an Implantable Antenna for Biotelemetry *International Journal of Scientific and Research Publications* **(7)** p 105-114

[5] Mandai B, Joseph L, Ebrahimizadeh J, Perez M D, Mitra D and Augustine R 2020 Low Profile Implantable Antenna for Fat Intra-Body Communication *The 14th European Conference on Antennas and Propagation (EuCAP)* p 1-4

[6] Das S and Mitra D 2018 A Compact Wideband Flexible Implantable Slot Antenna Design with Enhanced Gain *IEEE Transactions on Antennas and Propagations* **(66)** p 4309-4314

[7] Green R B and Topsakal E 2019 Biocompatible Antennas for Implantable Biosensor Systems *2019 International Workshop on Antenna Technology (iWAT)* p 70-72

[8] Yousaf m, Bashir Z, Amin Y and Akram A 2019 A Miniaturized Wideband Implantable Antenna for Telemetry Applications *2019 International Conference on Electrical, Communication, and Computer Engineering (ICECCE)* p 1-5

[9] Faisal F, Zada M, Ejaz A, Amin Y, Ullah S and Yoo H 2020 A Miniaturized Dual-Band Implantable Antenna System for Medical Applications *IEEE Transactions on Antennas and Propagations* **(68)** p 1161-1165

[10] Ali A A, Hamdan A, Kamardin K and Yamada Y 2019 Synthesis of Composite Material with Aerogel for Patch Antenna Substrate *The 2019 Malaysia-Japan Workshop on Radio Technology (MJWRT)* p 1-3

[11] Paul L C, Hosain M S, Sarker S, Prio M K, Morshed M and Sarkar A K 2015 The Effect of Changing Substrate Material and Thickness on the Performance of Inset Feed Microstrip Patch Antenna *American Journal of Networks and Communications* **(4)3** p 54-58

Acknowledgments
The authors would like to thank the Ministry of Higher Education (MOHE), Japan International Cooperation Agency (JICA), Communication Systems and Network (CSN i-Kohza) Research Group, Malaysia-Japan International Institute of Technology (MJIIT), and Universiti Teknologi Malaysia (UTM), Kuala Lumpur. This research is supported by JICA Joint Research Grant 2021.

Journal of Physics: Conference Series 2250 (2022) 012006 doi:10.1088/1742-6596/2250/1/012006

Focal Region Ray Tracing for Slanted Incident Rays on Spherical Reflector Antenna

Ayuni Afiqah Arjunaidi[1], Yoshihide Yamada[1], Kamilia Kamardin[1]

[1] Malaysia-Japan International Institute of Technology, Universiti Teknologi Malaysia, Jalan Sultan Yahya Petra, 54100 Kuala Lumpur

ayuniafiqah@graduate.utm.my

Abstract. Multi beam antenna system is required at base station for fifth-generation (5G) mobile communication and also to cover large portions of field of view for the satellite antenna. A spherical reflector antenna is promising to give high gain directivity of each independent beams for its multiple beam capabilities. The study of the feed elements system is carried out by analysing the focus of slanted incident rays in receiving mode from 0° to 30° at step angle 10°. Simple analytical equations are derived to find the incident and reflected rays. This focal region of spherical reflector is studied in MATLAB program. Accuracy of the MATLAB results are validated by examining the focal region using an electromagnetic simulator, FEKO. The locus of feed of spherical reflector antenna for certain angles of the incident rays are presented in this study.

1. Introduction

The proposed millimeter waves in 5G have high bandwidth potential that can transmit large amount of data in shorter time. However, higher frequencies transmission has less penetrating capability and small coverage distance [1]. Thus, beamforming capability is required to support massive multiple-input multiple-output (MIMO) application in small cell radius at approximately 200 cm [2]. Multiple beams antenna can serve a large number of users simultaneously and support multicast transmissions in broadcasting services [3]. Hence, providing spherical reflector antenna as promising candidate with high directional beam and wide-angle scanning performances due to its symmetrical surface [4].

The primary issue with spherical reflector is the spherical aberration, particularly in large aperture reflector. Based on the previous configuration by Ishimaru, et. al in [5], double spherical reflectors antenna that has the same center point for main and sub-spherical had shown that spherical aberration can reduce antenna gain. A continuation of the research is carried out in [6] to clarify the caustic positions of spherical reflector. Thereby, analysis on focal region ray tracing is useful to study the aberration effect and determination of feed position is important for multi beam radiation pattern.

In this paper, focal region ray tracing of spherical reflector for slanted incident rays at certain angles, from 10° to 30° is studied and illustrated by MATLAB program. The accuracy of MATLAB results are validated by using an electromagnetic simulator, FEKO. The focal region ray tracing is developed in FEKO by employing Ray Launching – Geometrical Optics (RL-GO) solver for the incident and reflected rays in receiving mode condition. The off axis focal region is illustrated in MATLAB and the intersection coordinates of the desired reflected rays are found to determine the adequate feed position.

2. General Geometry of Spherical Reflector

The geometry of spherical reflector is rotationally symmetrical around the axis of symmetry, y-axis as shown in figure 1. While z-axis corresponds to the radial length of the spherical reflector. The incident rays are excited in receiving mode and measured from a longitudinal distance at spherical center, C_M. The ray path of slanted incident rays (black) at given angle, $\Delta\varphi$ is illustrated with respect to its horizontal incident condition (blue) with total reflection, φ at given radial length. The parameters of the spherical reflector are shown in table 1.

Table 1. Spherical Reflector Antenna Configuration.

Antenna Parameter	Dimension
C_M, r_M	200 mm
φ_o	80°
$\Delta\varphi$	0°, 10°, 20°, 30°

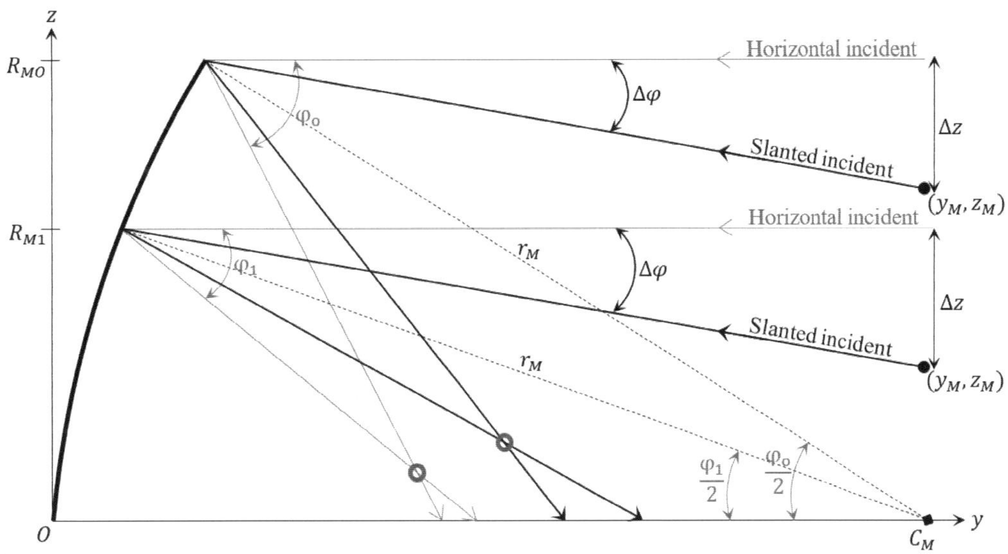

Figure 1. Two-dimensional (2D) geometry of spherical reflector and ray path of parallel incident rays.

3. Focal Region Ray Tracing in MATLAB

In order to find the focal region of the reflected rays, simple equations for slanted incident ray at given angle, $\Delta\varphi$ subjected to its horizontal incident ray, is derived below. Two conditions of horizontal incident ray at φ_o and φ_1 are shown as the examples and its caustic point is indicated by ⭘ mark. The cartesian coordinate for slanted incident ray (y_M, z_M) is expressed in the equations (1) to (3) below.

$$y_M = C_M \tag{1}$$

φ is presented by the total reflection angle of its horizontal condition such as φ_o or φ_1.

$$z_M = R_M - \Delta z \tag{2}$$

$$z_M = \left(r_M \sin\left(\frac{\varphi}{2}\right)\right) - \left(r_M \cos\left(\frac{\varphi}{2}\right) \tan(\Delta\varphi)\right). \tag{3}$$

Referring to the law of reflection as the fundamental of ray tracing concept, the tangent line is erected to a normal point where the reflection line is drawn; sphere radius r_M from centre C_M. The incident ray is reflected from the spherical reflector at its radial position R_M, either at R_{MO} or R_{M1}.

The direction along which these incident rays will be reflected are then identified in MATLAB by deriving equations of lines, $y = mx + c$. The MATLAB results of equation (1) and (3) are shown in the following figure 2.

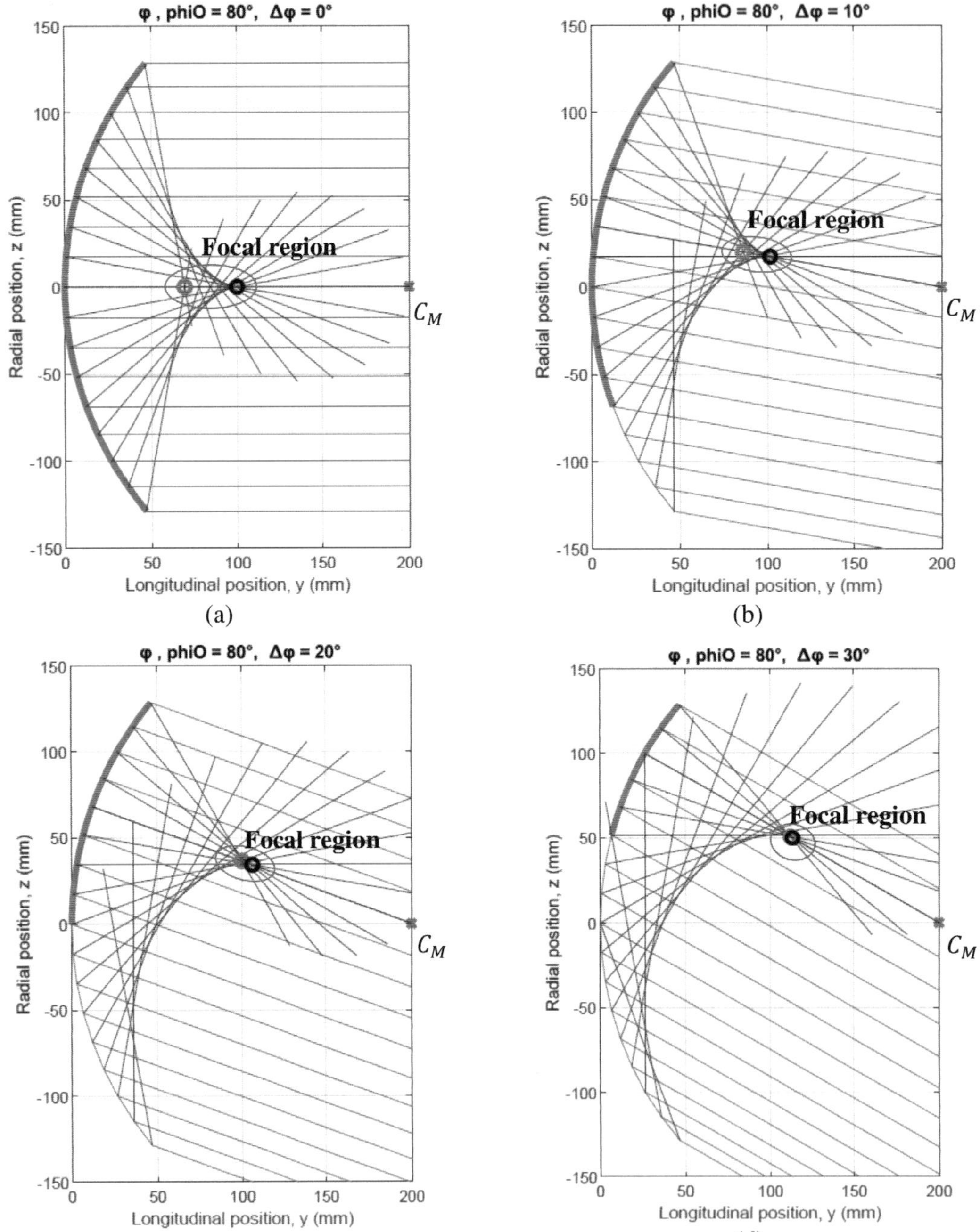

Figure 2. Focal region ray tracing for $\Delta\varphi$ equal to (a) 0°, (b) 10°, (c) 20°, and (d) 30° respectively.

From figure 2, the caustic curve for different angle of $\Delta\varphi$ is different. With the increasing angle of $\Delta\varphi$, the reflected rays unveil a larger curvature of caustic curve on the lower part (negative z-axis) of the spherical reflector. The reflected rays' concentration becomes asymmetric around the y-axis if $\Delta\varphi$ is more than 0°. Hence, at bigger angle of $\Delta\varphi$, the focal region becomes smaller by considering only few rays that are reflected towards the same direction of focusing area. Therefore, only fewer coordinate of feeds to select from smaller focal region. The thick red line indicates the contribution curvature of the spherical reflector focal region. Two possible coordinate of feeds in figure 2(d) are closer with each other compared to 2(a), 2(b) and 2(c) for $\Delta\varphi$ equal to 0°, 10° and 20° respectively that have larger focal region. These are obtained from two intersection points of reflected rays at ○ mark in the focal region by using `polyxpoly` syntax.

4. Focal Region Settings in FEKO

In FEKO, focal region ray tracing of spherical reflector is conducted by using RL-GO ray launching settings. Ray contributions are examined for direct and reflected conditions. The spherical reflector is modelled by importing coordinates from MATLAB structure through polyline method. Next, a free space region medium for the spherical reflector with length equal to 200 mm is created as the boundary of the ray tracing as shown in figure 3. These geometries are set to solve with RL-GO solution under properties option. A plane wave source is added into the configuration by adjusting the setting for loop over multiple directions (θ) at given $\Delta\varphi$ as shown in figure 4.

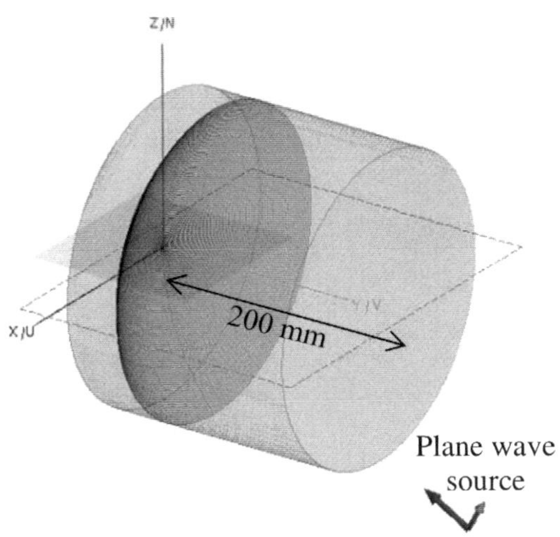

Figure 3. 3D geometry of spherical reflector

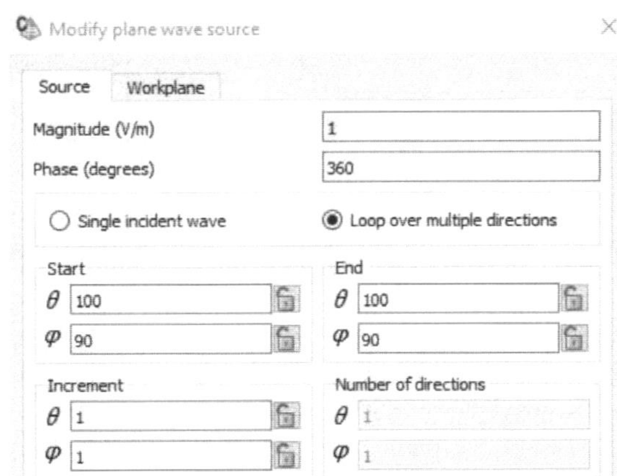

Figure 4. Plane wave source setting for focal region ray trace at θ either 100°, 110°, or 120° for $\Delta\varphi$ equal to 10°, 20°, or 30° respectively.

The plane wave excitation is significant to produce the incident rays and then will be reflected inwards by the spherical reflector surface. A focal region can be seen from the collection of intersection reflected rays in the air region. Plane wave direction from few angles, $\Delta\varphi$, are shown in figure 5(a), 5(b) and 5(c).

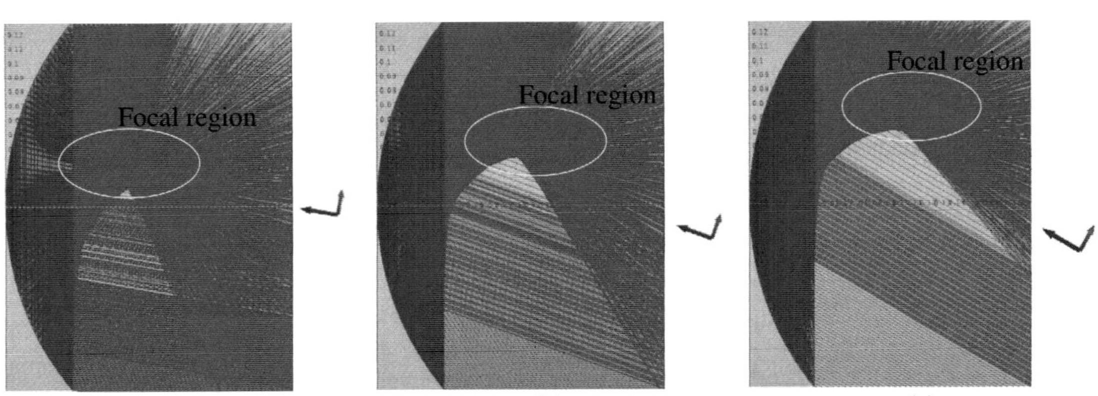

(a) (b) (c)

Figure 5. Shifted plane wave at $\Delta\varphi$ equal to (a) 10°, (b) 20°, and (c) 30° respectively.

Figure 5 clearly shows the upward movement of the focal region from 10° to 30°. Both MATLAB and FEKO ray tracing results are compatible and reflected rays are intersected at the same locations. Measure distance tool in POSTFEKO is used to confirm the distance of intersection points in MATLAB to choose the potential feed positions.

5. Locus of feed

From focal region analysis, various possibilities of feed position can be determined for multiple beam operations. Two approximate feed positions from the focal region are determined for the multi beam mechanism as shown in table 2 and figure 6.

Table 2. Coordinate of Feed Positions

	Feed position (1)		Feed position (2)	
$\Delta\varphi$ (°)	y-coordinate (mm)	z-coordinate (mm)	y-coordinate (mm)	z-coordinate (mm)
0	69.45	0.00	99.99	0.00
10	86.28	20.05	101.51	17.36
20	100.00	36.39	106.02	34.20
30	112.06	50.77	113.39	50.00

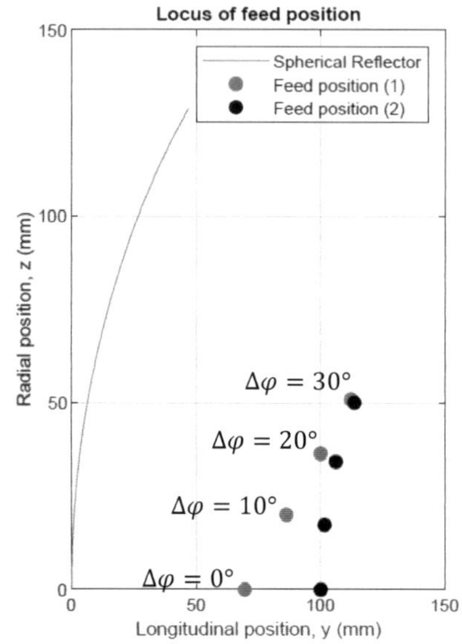

Figure 6. Locus of feed positions with spherical reflector

6. Conclusion

Focal region ray tracing at slanted incident angle is derived and solved by MATLAB program which later is proven by FEKO simulation. However, there are many possibilities of feed position at lower shifted angle of plane wave due to larger focal region. Since the spherical aberration is moving, the feed determination is only applicable for the upper part of the spherical reflector, which then an offset spherical reflector is considered for good multi beam performance. Future work is needed to ensure good multibeam radiation patterns from transmitting mode of the feeder antenna on offset spherical reflector.

References

[1] Xingdong, P., et al., *Design and implementation of an active multibeam antenna system with 64 RF channels and 256 antenna elements for massive MIMO application in 5G wireless communications.* China Communications, 2014. 11(11): p. 16-23.

[2] Chataut, R. and R. Akl, *Massive MIMO Systems for 5G and beyond Networks—Overview, Recent Trends, Challenges, and Future Research Direction.* Sensors, 2020. 20(10).

[3] Rinaldi, F., et al., *Broadcasting Services Over 5G NR Enabled Multi-Beam Non-Terrestrial Networks.* IEEE Transactions on Broadcasting, 2021. 67(1): p. 33-45.

[4] Volakis, J., *Antenna Engineering Handbook.* Vol. 23. 2007.

[5] Ishimaru, A., I. Sreenivasiah, and V. Wong, *Double spherical cassegrain reflector antennas.* IEEE Transactions on Antennas and Propagation, 1973. 21(6): p. 774-780.

[6] Quzwain, K., et al. *Caustic Analysis of Reflected Rays from a Spherical Reflector Antenna.* in *2020 IEEE International RF and Microwave Conference (RFM).* 2020.

Acknowledgments

The authors wish to acknowledge and thank Universiti Teknologi Malaysia (UTM) and Ministry of Higher Education for providing the Research Grant (FRGS/1/2018/TK04/UTM/01/4).

Journal of Physics: Conference Series 2250 (2022) 012007 doi:10.1088/1742-6596/2250/1/012007

Design Strategies to Overcome EMI/EMC Issues of Airborne Receivers

Mahadev Sarkar[1] and Raghu B R[2]

[1]Product Development and Innovation Centre, Bharat Electronics Ltd
Prof. U R Rao Road, Jalahalli, Bangalore 560013, India
[2]MWSC, Bharat Electronics Limited Bangalore 560045, India
Email: mahadevsarkar@bel.co.in

Abstract. This article describes various housing design strategies of a complex receiver system to overcome EMI /EMC challenges for airborne system applications. Isolation of leakage energy between various RF/DC-DC converters/Regulators /pulsing blocks are the key target of this design to keep conductive and radiative emission minimum. Hence, complex receiver design has been segregated into various modules. Block level design is carried out using 3D EM simulator to assess the leakage. Then they are stacked together very tactfully within the specified area without compromising inter-channel /module interference between them and the form factor respectively. The super compact stacks are further channelized for RF, DC supply, controls etc. and protected them from any kind of interferences. Low cost flexi interconnects are made to interface various stacks to arrest unwanted high/low frequency leakages. Finally, receiver module is developed and tested over temperature and EMI/EMC successfully to establish the methodology.

1. Introduction

Ultra-wideband microwave receiver system plays an important role in contemporary passive seeker development where, frequency and wavelength signatures are necessary to track down an unknown threat(s) [1–2]. This design involves numerous pulse/data generations circuits which includes DC-DC converters /regulators, sequencing, block respectively. These circuits are producing huge number of voltage /current pulses which having sharp rise/fall time characteristics. These fast changing current/voltage pulses are one of the main reasons of EMI generation. Hence, EMI/EMC (Electro-Magnetic Interference/ Electro Magnetic Compatibility) tests become a challenging task for such airborne application systems [3–4]. Extensive works have been published over the mitigation of EMI/EMC issues starting from circuit level design to system level implementations [5–16]. A topological network based model is proposed to carry out system level EMC analysis in [4]. Model analysis/visualization and mitigation of parasitic coupling paths between various modules like RF–digital, C–L–C filter circuits, flyover QSFP Connectors, switching power converter, high density connectors interface etc. have been discussed in [6–11]. This proposed analysis and modelling are limited to either a single board or within a single metallic enclosure. This article explains mixed EM coupling paths between RF-digital stack, Digital-pulsing-sequencing board, RF-RF power stacks/ channels, inter-stack RF connector-power supply connector-high speed data transfer connector, etc. The aim of this design is to keep conductive and radiative emissions is at a minimum level. Therefore, this complex design is segregated into three major design chunks like, RF subsystem design, Digital Processing and Power supply sequencing module design respectively. Then each block are packed into one or multiple physically separable mechanical housing called stack and provided appropriate interconnects to arrest any kind of signal leakages. RF block is further divided into various modules. Total of four portable/physically separable stacks have been planned to accommodate all these design modules. Three stacks have been planned and used for RF design implementation. A Digital signal processing block has been implemented in the fourth stack. DC Supply and control block shares first and third stacks respectively. Then flexi interconnects have been used to interface various stacks. Processor board with DC supply modules are also integrated into the receiver (RF unit) as another stack and all are isolated from each other using inner and inter-stack EMI gaskets, absorbers, metallic mesh, etc. to qualify temperature, vibration and emission tests respectively.

2. Design Strategies and Planning

This complex receiver design follows in figure 1 and steps to overcome EMI/EMC related issues. Thereby the whole design has been divided into several units like; high power RF/LO generation module, low power RF blocks, passive sections, FPGA based digital signal processing sections and linear regulator-OPAMP based circuits [11]. High power RF/LO generation modules produce power from -55dBm to +20dBm (max.) fixed power over a wide frequency range thereby this stack generates excessive heat. More surface area is required to dissipate such excessive heat and that is why the bottom most stack has been planned for high power generation as it is touched to the outer chassis of the seeker through the bottom cover of the receiver. Medium, low and passive RF units have been accommodated in second and third stacks. Half of the area of forth stack have been used for high speed pulsing and digital processing circuits. The remaining area of the stack has been used for high power supply and control generation circuits as this stack is also touched by the outer chassis of the seeker through the top cover of the receiver. Supply sequencing and a few control generation circuits have also been accommodated at the first stack also. RF and DC supply interconnections are carried out very carefully. As mentioned above, high power RF, digital signals, clocks and control pulses are running very close to each other thereby mixing of different signals are unavoidable.

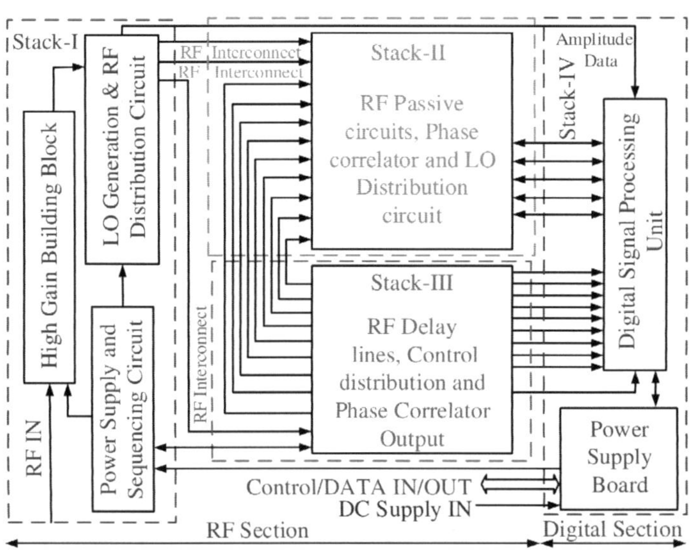

Figure 1. Complete Stack details and its interconnectivity of digital processor.

Proper channelling is the first step to overcoming signal mixing. Henceforth, proper channelling of all different level power signals, clocks, control signal, and shielding arrangements have been planned [15–16]. Finally, four modules are developed and tested over -40°C to +71 °C (ESS) pre and post temperature vibrations and EMI/EMC successfully to establish the methodology. The complete stacks details with all interconnect are explained in figure 1. Here stack-I accommodates high gain building blocks, LO generation unit, input amplitude tracking unit and medium level power supply and sequencing unit respectively. High gain building block gets very low level (≈ -

55dBm) input RF power from the external world and converts into high level output. LO generation block further rectifies, equalizes the input RF power and makes a fixed output power suitable for phase-correlators [10]. The power supply and sequencing unit get input supplies and controls signals from Stack-IV as shown in figure 1 respectively. Stack-II and –III accommodate all complex RF modules. These two stacks are the most compact and several interconnects are placed between them. Thereby it is the most vulnerable to produce unwanted signals either in terms of conductive or radiative emissions respectively.

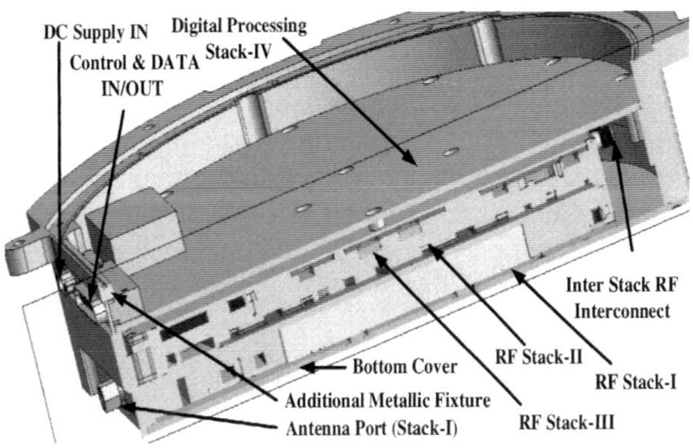

Figure 2. Complete Stack details of receiver

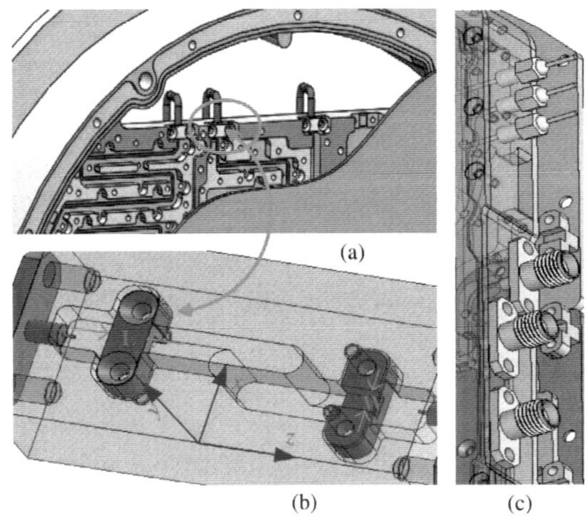

Figure 3. (a) RF interconnects (Stack –I to –II & –III) flexi cable interconnects (b) T-junc. to hold the flexi cable, (c) Connection of Stack –II & –III RF flexi cable from Stack –I.

3. Design Implementations

Final stack up of receiver integrated with high speed digital signal processing board and mechanism of RF interconnects are shown in figure 2. Figure 3 shows the overall stack pile up, it depicts that, Stack –I and –IV has an external antenna port, DC supply and high speed data in/out interface respectively to communicate the external world. Super compact Stack–II and –III have multiple RF/high frequency, DC and high speed data interface which can easily interact with Stack–I and –IV circuitry and then will directly interact with outer world through interconnects. Figure 3.a shows how stacks are connected using flexi cables. Low cost inter-stack RF interconnects have special type of assembly mechanism to reduce the accommodation area in Stack –I. Here special T junctions have been used to fix the flexible cables and are shown in figure 3.b. thereby from the above figure it is clearly evident that high frequency RF has fair possibility to leak through T junctions. Unwanted radiation from a small aperture has severe effect over RF performance and it directly contributes to EMI/EMC hazard. Another side of the flexi cables use a connector to assemble with Stack–II and –III to ease the post assembly troubleshooting and stage wise module testing. Power supply and high speed digital signal processing board have been divided into major three sectors like high power DC-DC converter with isolated output, multi-out regulators, low -medium power linear regulators, pulsing, sequencing board and high speed digital signal processing segment respectively. Linear regulators, pulsing, sequencing board is kept at Stack –I with high power RF generation modules. Other two segments are implemented in Stack –IV as mentioned above. Although linear regulator and pulsing board is simpler as compared to DC and high speed board design (in Stack –IV), yet utmost care has been taken to design these boards with proper grounding. Due to switching nature of DC-DC converters, EMC is critical of these converters to ensure its own action as well as the safe and secure operation of surrounding other electronic components. General rules are followed to design this board like; reduction of loop inductance, input/output filtering, layout optimization technique, etc. respectively. Here analogue and digital grounds are separated and similar kind of circuits is also grouped together with proper layer definitions. Power planes for high current carrying supplies are kept in separate layers with enough ground spacing. High speed data/control carrying traces are designed in strip line form to keep radiation effect minimum. Power and signal integrity have been checked for all the traces to avoid any kind of unwanted phase. Layout optimization technique follows the below steps:

a. Decoupling capacitors placement: all the decoupling capacitors are kept as close as possible to the devices. Within those capacitors again, small value capacitors are kept nearest to the device and then higher value capacitors follow the standard placement rules.

b. Active gate/Transistors placement: in the pulsing and sequencing circuits, transistors are placed at a minimum distance from each other to keep minimum path inductance.

c. Live and ground via placement: live/ground vias have been placed after carefully examining the current path in the loop, trace to trace distances, layer to layer buried via locations etc.

d. Dimension of the phase plane is kept small enough to minimize electromagnetic coupling.

e. Device package selection: package style has a role to influence the loop inductance, thus package style has been optimized to reduce loop inductance.

These optimization techniques have been followed to design all the supply and high speed digital signal processing cards, respectively.

4. Design Realization

All hermetic modules are sealed after ensuring proper EMI gasket/absorbers protection. Then it becomes ready for second level module integration. First level testing is conducted by using individual power supplies and external controls respectively. The second stage of assembly/ integration is basically stack level integration. Unlike level-1, level-2 (stack level) testing is conducted by taking all power supplies and controls from Stack–I & –IV. A sample of compact RF packaging with high/low power LO/RF sections are shown in figure 3. High power LO and RF sections are separated by metallic walls. There are a few (wherever required) extra metallic extrudes are shown within the enclosed dotted area between inner metallic cover placement areas. These are required to stop unwanted RF leakage from one section to another section. It is known that there will always be micron level gaps between well finished housing and inner covers. A few samples of absorber clad metallic inner covers and RF-DC cross over beneath the hermetic module are shown in figure 4.

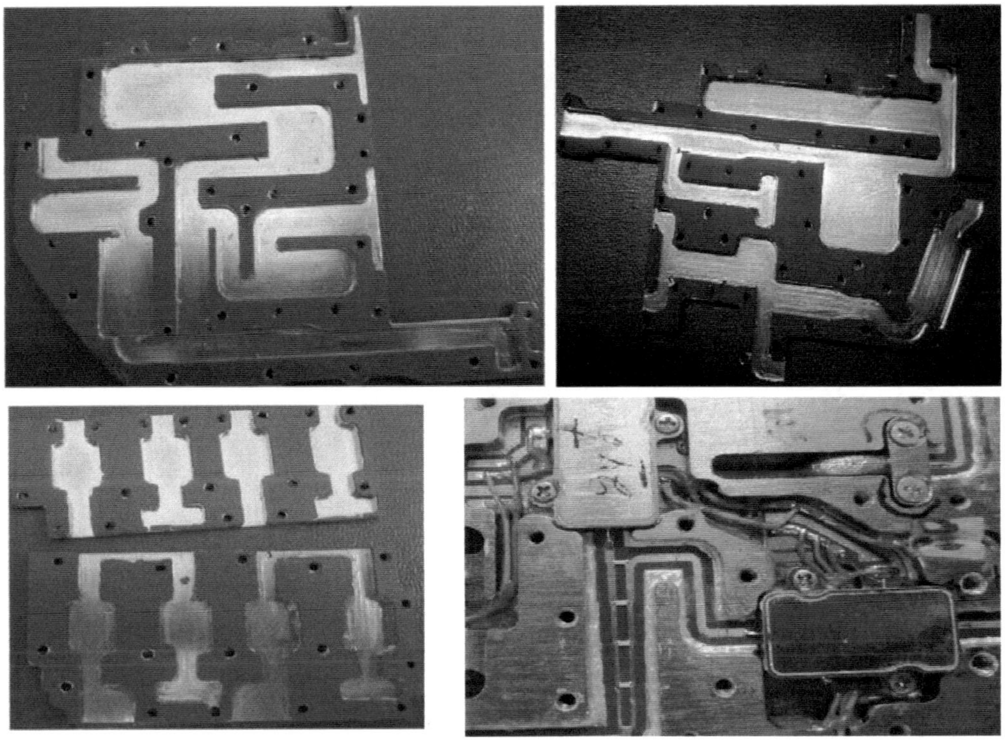

Figure 4. Individual absorber-clad metallic inner covers to shield any unwanted emissions to adjacent channels & RF DC crossover under hermetic module

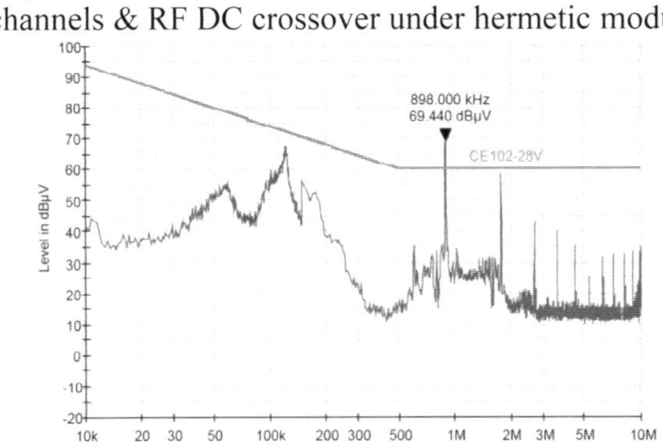

Figure 5. CE-102 measured on +28V LISN line of receiver before filtering.

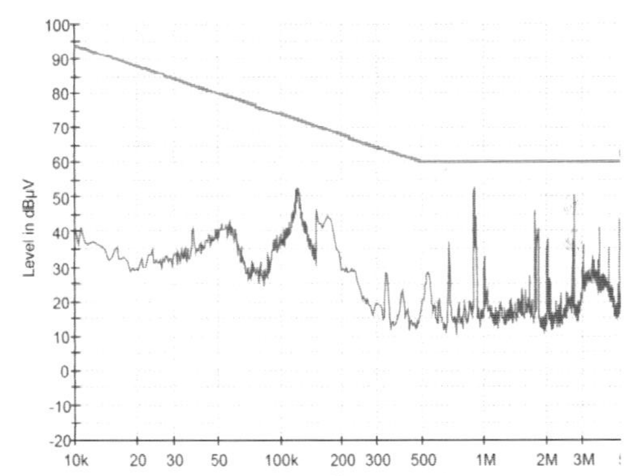

Figure 6. CE-102 measured on +28V LISN line of receiver after filtering.

These inner covers have module specific grooving for suitable RF performance and to accommodate absorbers/EMI gaskets. As indicated above, these covers are used multiple times onto a specific area to arrest any kind of EM emission from closely spaced channels. RF interconnections between Stack –I and Stack –II, –III are shown here. Now EMI/EMC tests have been performed under ambient temperature of 25°± 3°C and 45 – 70% humidity respectively. All units have under gone conducted emission (CE102), radiated emission (RE102), conducted susceptibility (CS101, CS114, CS115 & SC116) radiated susceptibility (RS103). During the test EUT (equipment under test) was powered on with +28V supply and a signal of frequency from 1 – 10GHz. Measured results of CE-102 on 28V LISN (line impedance stabilizing network) line is shown in figure 5. Root cause analysis of 900KHz signal has been carried out and it showed that DC-DC converter has switching frequency of 900KHz. Filtering of the DC-DC converter is insufficient and thereby switching frequency is getting leaked through 28V supply lines. Now faulty EMI filter and some decoupling capacitors are also replaced and the switching frequency emission problem is solved and shown in figure 6. Radiative emission (RE) test has been performed over 2MHz to 18GHz band aligning antenna at vertical (V-pole) and horizontal (H-pole) polarization configurations respectively and are shown in figure 6 and -7 respectively.

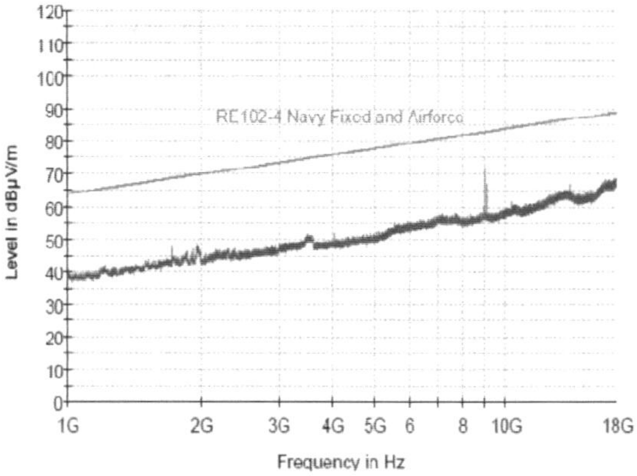

Figure 7. RE-102 measured from 1GHz – 18GHz for vertical polarization

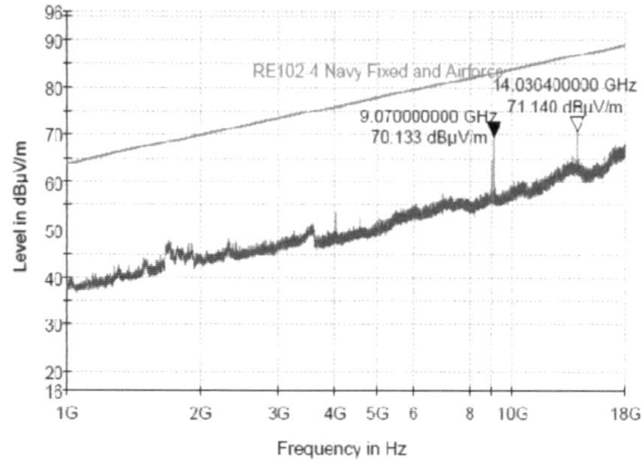

Figure 8. RE-102 measured from 1GHz – 18GHz for horizontal polarization

5. Conclusion

EMI/EMC suppression challenges have been discussed in this paper. RF and digital circuit stacking plans and executions have been discussed in details. The effect of emissions on system performance are discussed in this article and based on calculation and the whole module design scheme is split into numerous modules. Electromagnetic emission from small gap is anticipated in this article. Various strategies have been implemented to arrest unwanted emission. Module design plan and realization stages are ensured phase by phase to make it effective. DC supply board has to be redesigned as supply filtering was not done correctly. DC board layout has to be optimized with better filtering. Thus, it can be concluded that the module design and execution plans are successful. RF modules placed in various stacks can be multilayer designs thus a number of RF stacks can be reduced or overall module size can further be reduced. New EMI/EMC challenges may arise by doing so and those challenges have to be mitigated with proper planning.

References

[1] M. Sarkar, V. Kumar, R. Sivakumar, S. Jhariya, B. R. Raghu and K. S. Sudheesh, "Some Investigations on Direct Substrate Attachment Process for Wideband Instantaneous Frequency Measurement Receiver Development" IEEE Transactions on Components, Packaging and Manufacturing Technology, vol. 10. no. 4, pp. 704-716, April, 2020 doi:10.1109/TCPMT. 2020.2966709.

[2] M. Sarkar, "Sharp Rejection Wideband Band Pass Filter in Suspended Substrate Stripline Realization", IEEE International Conference for Convergence Technology (I2CT), pp. 1–4, March, 2019, doi: 10.1109/I2CT4561.2019.9033814.

[3] Y. S. Cao, Y. Yansheng, L. Jiang, A. E. Ruehli, J. Fan and J. L. Drewniak, "Quantifying EMI: A Methodology for Determining and Quantifyinh Radiation for Practical Design Guidelines" IEEE Transactions on Electromagnetic Compatibility, vol. 59, no. 5, pp. 1424 -1423, June. 2018. doi 10.1109/TEMC.2017.2677199.

[4] A. Bendicks, T. Dorlemann, S. Frei, N. Hees and M. Wiegand, "Active EMI Reduction of Stationary Clocked Systems by Adapted Harminics Cancellation" IEEE Transactions on Electromagnetic Compatibility, vol. 61, no. 4, pp. 998 -1006, Aug. 2019. doi 10.1109/TEMC.2019.2854965.

[5] M. Li, Y. Wen, and G. Wang, "EMC Analysis of Train Control System based on Network", IEEE Letters on Electromagnetic Compatibility Practice and Applications, vol. 2, no. 4, pp. 101 -105, June. 2020. doi 10.1109/LEMCPA.2020 .3014735.

[6] L. Ren and J. Fan, "Modeling Electromagnetic Field Coupling Through Apertures for Radio-Frequency Interface Applications", IEEE Transactions on Electromagnetic Compatibility, vol. 57, no. 5, pp. 1329 -1336, Oct. 2015. doi 10.1109/TEMC.2015.2424248.

[7] Y. Murata, K. Takahashi, T. Kanamoto and M. Kubota, "Analysis of Parasitic Couplings in EMI Filters and Coupling Reduction Methods," IEEE Transactions on Power Electronics, vol. 59, no. 6, pp. 1880 -1886, Maech. 2017. doi 10.1109 /TPEL.2017.2690457.

[8] A. Talebzadeh, K. Vuppunutala, K. Koo, H. Li, J. Nadolny, Q. Liu, J. Li, K. Ghosh, P.C. Sochoux, V. Khilkevich, J. L. Drewniak and D. Pommerenke, "Coupling Path Visualization and EMI Mitigation for Flyover QSFP Connectors", ", IEEE Transactions on Electromagnetic Compatibility, vol. 62, no. 4, pp. 1037 -1044, Aug. 2020. Doi 10.1109 /TEMC.2019.2943290.

[9] M. Jin, M. Weiming, P. Qijun, K. Jun, Z. Lei and Z. Zhihua, "Identification of Essential Coupling Path Models for Conducted EMI Prediction in Switching Power Converters" IEEE Transactions on Power Electronics, vol.21, no. 6, pp. 1795-1803, Nov. 2006, doi 10.1109 /TPEL.2006.882905.

[10] Sarkar M, Sivakumar R and Manjunath R., "Highly Selective Cavity based Band Pass Filter Using Asymmetric Double Quadruplets" IEEE Internation Conference on Microwave and Optical Communication Engineering (ICMOCE), pp. 56-59, December, 2015

[11] C. Wu, H. Kim, J. He, N. Erickson, S.Chao, D.Kim, Y. Hur, D. Pommerenke and J Fan "Analysis and Modelling of Conducted EMI From an AC–DC Power Supply in LED TV up to 1 MHz", IEEE Transactions on Electromagnetic Compatibility, vol. 61, no. 6, pp. 2050 -2059, Nov. 2019. doi 10.1109/TEMC.2019.2954360.

[12] M. Sarkar, et al, "Performance Stabilization of Steep Gain Gradient class Ultra-wideband Direct Conversion IFM Receiver of Homodyning based Systems" International Journal of Applied Science-EPH, Vol. 3 Issue. 4, pp. 27 – 31, April, 2017.

[13] M. Sarkar,G. anand, R. Sivakumar, "Light weight Active 1–18 GHz Ultra High Speed SPDT Switch for Radar Warning Receiver Systems," IEEE MTT-S International Microwave and RF Conference (IMaRC), June, 2020, doi:10.1109/IMaRC2019.9116744.

[14] M. Sarkar, G. anand, R. Sivakumar, "Wideband Active Switch for Electronic Warfare System Applications" ASTES journal, vol. 5, no. 6, pp. 1312 – 1321, Dec. 2020, doi.: https://dx.doi.org/10.25046/aj056156.

[15] M. Sarkar, S. Jhariya and R. Sivakumar, "Correction of Nonunifirm Attenuation of Voltage Contrilled Attenuator under Harsh Environment by Process Improvement" IEEE Transactions on Components, Packaging and Manufacturing Technology, vol. 9, Issue 8, pp. 1632-1642, Aug, 2019, doi 10.1109/TCPMT.2019.2929093.

[16] M. Sarkar, B. R. Raghu and K. S. Sudeesh, "Evaluation of Doubled-sided Substrate Attachment Technique for Round Shaped Large Metallic Housing under Airborne System Development" 2020 IEEE 5th International Conference on Computing Communication and Automation (ICCCA), pp. 191-194, Nov. 2020, doi: 10.1109/ICCCA49541.2020.9250 792.

Acknowledgments
Authors gratefully thank Shri. Suresh Kumar KV, Head, PDIC, for providing facility and inspiring to publish this work. Authors also thank to Shri Debasish Das, Harikrishna M V and Sivakumar R, for their valuable suggestions. Author is grateful to MWSC and product design group for giving their support. He also thanks all others who were involved directly or indirectly in this work

Detection and Classification of Target's Speed and Size Using LTE-Based Passive Forward Scattering Radar

N H Abdul Aziz[1*], M F Abdul Hadi[1], N H Abd Rahman[1], A J Alias[2], and M R J AL-Hiealy[2].

[1]Antenna Research Centre, School of Electrical Engineering, College of Engineering, Universiti Teknologi MARA, 40450 Shah Alam, Selangor, Malaysia.
[2]University Malaysia of Computer Science & Engineering (UNIMY), 63000 Cyberjaya, Selangor, Malaysia.

Email: noor4083@uitm.edu.my

Abstract. Radar is a system that uses radio waves to detect any objects that pass through the range of its system. The radar system must have two main components which are the transmitter and the receiver. The passive forward scattering radar is being used to detect the target's speed and size. This passive radar uses the cell tower of Long-Term Evolution (LTE) as a transmitter and will illuminate any moving targets that pass through the system. The passive radar receiver which can be remotely placed will detect the scattered radiation of the target. The detection of the target is in a waveform called the Doppler signature. From the Doppler signature, the target's speed and size can be classified according to its group. In this research, a different target has a different Doppler signature because every target's body has a different body build. The differences in Doppler signature also can be detected when the target moves at different speeds. The study of different speed targets will relate to the relationship between the time and frequency theory. The data collection of Doppler signature is collected in three different places which are Universiti Teknologi MARA (UiTM, Shah Alam), Kampung Sungai Kajang (Tanjung Karang), and Malaysia Agro Exposition Park Serdang (MAEPS) (Serdang). In conclusion, the LTE-based passive forward scattering radar system is applicable to detect and classify every target with different speed and size which is useful to the country's border safety and security system.

1. Introduction

Forward Scatter Radar is a type of bistatic radar. The signal received from the target detection is due to the direct path signal shadowing the target body. For the passive forward scattering radar, the receiver can remotely change location because the receiver and the transmitter are separated at a considerable distance. Since the system can only receive the signal and cannot produce any signal, this passive forward scattering radar can provide a number of advantages, such as enhanced radar cross-section, inherent detection ability of stealth target and reasonably low complexity design of system [1].
Passive radar is undetectable because it emits no radio energy and can be camouflages anywhere. It is beneficial to the military forces since the radar is difficult to locate and target [2]. This study focuses on the detection and classification of a target's speed and size using passive forward scattering radar. The Long-Term Evolution (LTE) signal was used as the transmitter for this system. The transmitter transmits a signal to detect the target and that signal will have a change in properties when it is received by the receiver [3]. This is called the Doppler Effect. The change in property of the signal is due to the scattering of the signal that collide with the target [4]. In this case, every speed and size produce a different Doppler signature and can be classified into its own cluster. In term of speed, the relationship of speed, distance

and time was used to detect and classify all of the collected Doppler signatures. The collected data was then processed using MATLAB software.
To classify the data into its group, the Principal Component Analysis (PCA) was used to calculate the percentage of performance on the detection at each experiment site. PCA can classify the data into its own band or group so that the pattern and the behaviour of the signal can be easily viewed [5] and further study. To calculate the percentage of performance, the method of K-nearest neighbour was used by identifying the nearest neighbouring data to determine which group the data belong to [6]. For this study, the three-nearest neighbours were used to calculate the percentage of performance.

2. Methodology

2.1. Vehicle's Speed Detection

For detecting the speed of a vehicle, a motorcycle was used as the target to travel pass through the radar system. This experiment needed two persons, a rider and a passenger, on the motorcycle to differentiate the size of the target. First, the motorcycle with one passenger needed to travel at three different speeds, which were 10 km/h, 20 km/h and 30 km/h in a straight line passing between the transmitter (Tx) and receiver (Rx) at a distance of 1 meter from the receiver. The experiment was done at 1.8 GHz LTE signal frequency at MAEPS (Serdang). The experiment was repeated with the same motorcycle, but two riders on the back to detect and classified the different in Doppler signature of different size of target. All the information about the experimental setup is shown in table 1.

Table 1. Information about the number of samples at MAEPS.

Vehicle's Speed (km/h)	Number of Samples	
	One Person	Two Person
10	30	33
20	33	33
30	32	33

2.2. Human's Speed Detection

For the detection human speed, two persons were needed as the moving target called Small Person and Big Person to collect the Doppler data signature. They needed to travel pass through the radar system at three different speeds, which were 1 km/h, 5 km/h, and 12 km/h. This experiment was conducted at UiTM and Tanjung Karang at a frequency of 1.8 GHz. All the information about the experiment at both places is tabulated in table 2.

Table 2. Information about the number of samples at UiTM and Tanjung Karang.

Human's Speed (km/h)	Number of Samples			
	UiTM		Tanjung Karang	
	Small Person	Big Person	Small Person	Big Person
1	73	75	42	41
5	75	72	42	42
12	61	75	42	42

2.3. Experiment Site

There were a few aspects that needed to be considered to find an experiment site where the detection and classification of target's speed and size using LTE-based Passive Forward Scattered Radar can be done. First thing that needed to be considered was the LTE cell tower that acted as the transmitter for this system. After getting the LTE cell tower location, check the signal strength of the LTE signal of the cell tower was checked to be at a frequency of 1.8 GHz using the spectrum analyser. Then, the location was checked to see if there was any blockage of signal so that the noise can be reduced. Another aspect that needed to be considered was the safety of the experiment site. After all of the aspects were considered, the horn antenna for the passive radar was set up at the location. The set up for the LTE-based Passive Forward Scattered Radar is shown in figure 1.

The experiment was set up at three different places, namely MAEPS, UiTM, and Tanjung Karang. In MAEPS, the specific place of the experiment site was in the latitude and longitude of (2.981°, 101.703°). The distance of the target from transmitter (Cell Tower) was about 186.77 m. The location of the experiment site is as shown in figure 2. At UiTM, the specific place of the experiment site was in 3.073°, 101.497°. The distance of the target from transmitter was about 151.48 m as shown in figure 3. In Tanjung Karang, the specific place of the experiment site was in 3.417°, 101.167°. The distance was about 845.69 m and as shown in figure 4.

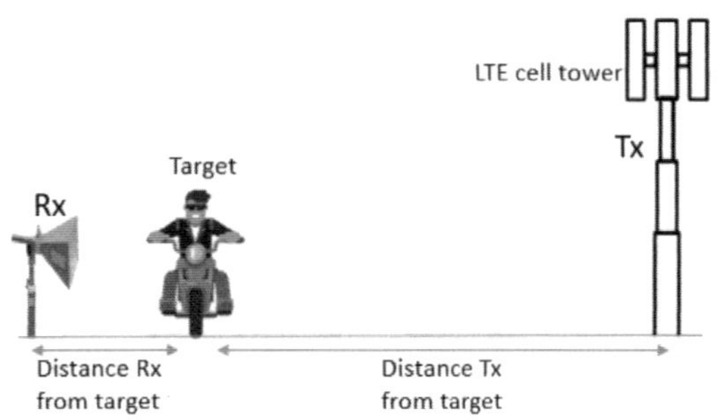

Figure 1. The illustration of experiment setup.

Figure 2. Experiment site at MAEPS.

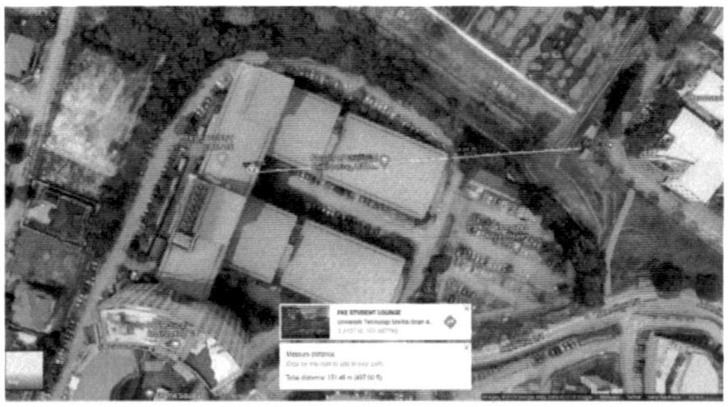

Figure 3. Experiment site at UiTM.

Figure 4. Experiment site at Tanjung Karang.

2.4. Signal Processing

The study on the detection and classification of target's speed and size started with the collection of Doppler data in each of the three experiment sites. The data collection recorded 20 seconds of data. Then, the data was segmented into 2 seconds. The 2 seconds data indicated the Doppler signature of the target. To reduce the noise in the Doppler signature of the target, the data needed to undergo a denoising process. Next, to analyze the data in term of frequency, the denoised data was converted from time domain to frequency domain using the Power Spectrum Density (PSD). After being converted into the frequency domain, the data was classified into their own clusters by Principal Component Analysis (PCA). Figure 5 shows the flowchart of the data processing for this study.

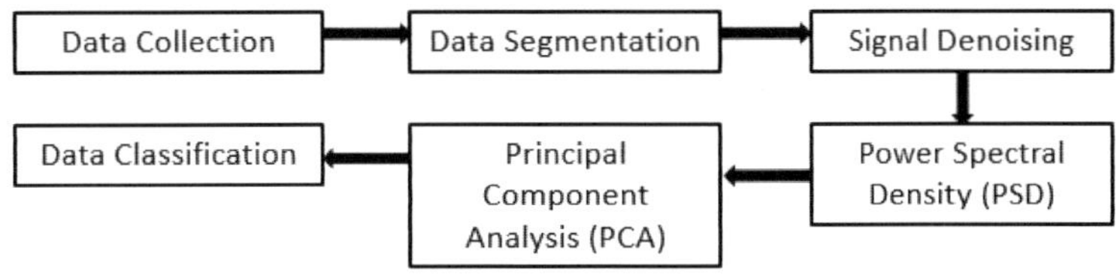

Figure 5. The flowchart of data analysis process.

3. Result And Discussion

3.1. Time Domain After Denoise Process

Figure 6, figure 7 and figure 8 display the result of time domain between the three different speeds at three different experimental sites. Each speed had its own Doppler signature. Figure 6 shows the result of time domain between the speed of 10 km/h, 20 km/h and 30 km/h of a person riding a motorcycle in MAEPS. From this result, as the speed increased, the time taken for the target to pass through the line between the transmitter and receiver tended to decrease. This was proven by the speed and distance relationship ($s = d/t$). There was also a difference in Doppler signature data in term of number of person on the motorcycle. When there was one person (red line) riding the motorcycle, who had a smaller body, the Doppler signature for the one person needed less time to pass through the line between the transmitter and receiver compared to two persons (blue line) riding the motorcycle.

The results for the next two experimental sites, which are UiTM and Tanjung Karang, tended to follow the trends. In figure 7 and figure 8, it shows the result of the time domain between the speed of 1 km/h, 5 km/h and 12 km/h of two different sized humans. Based on the results, the time taken for the targets to pass through the line between the transmitter and receiver tended to decrease as the speed increased. In term of body size, the Doppler data signature for the smaller human body (red line) was smaller compared to bigger human body (blue line). The smaller human body structure needed less time to finish travelling in the radar range.

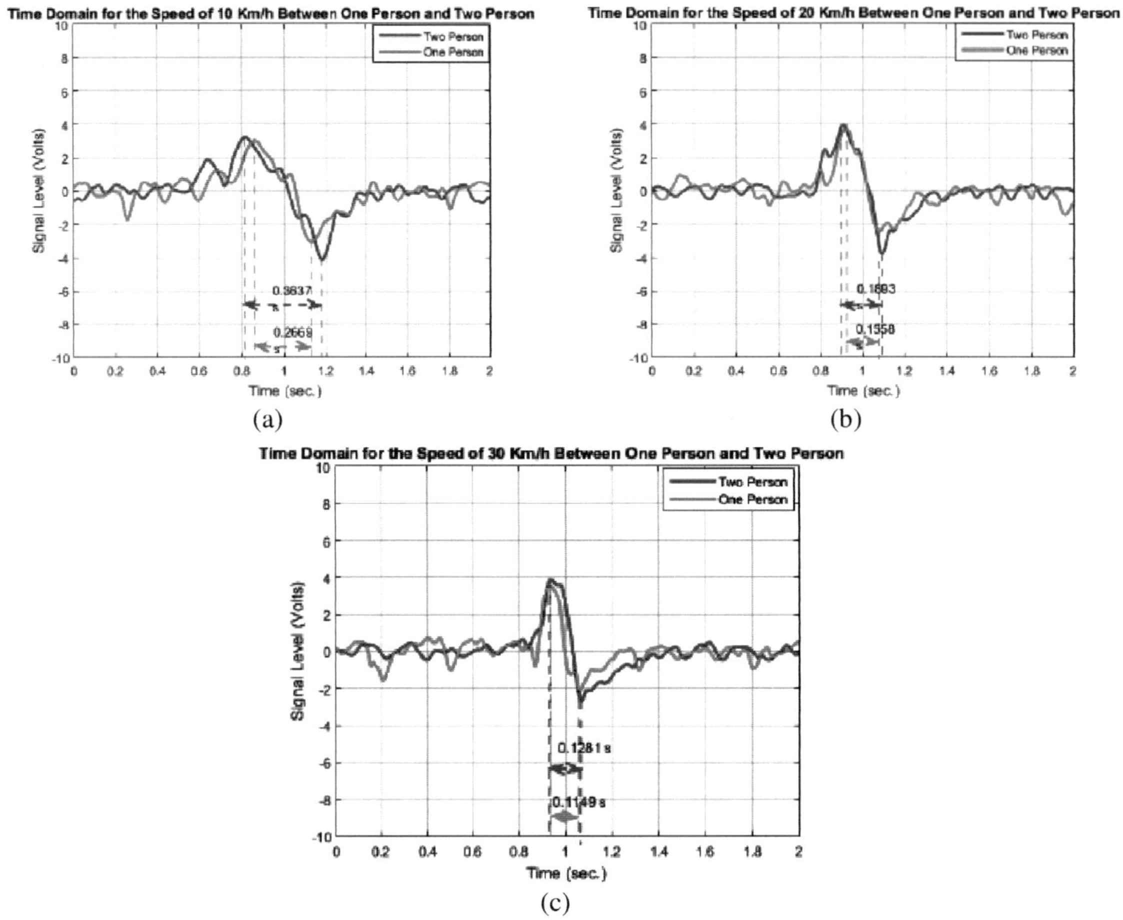

Figure 6. Time Domain of 1.8 GHz signal at MAEPS for the three different speeds of one person and two persons on motorcycle (a) 10 km/h, (b) 20 km/h and (c) 30 km/h.

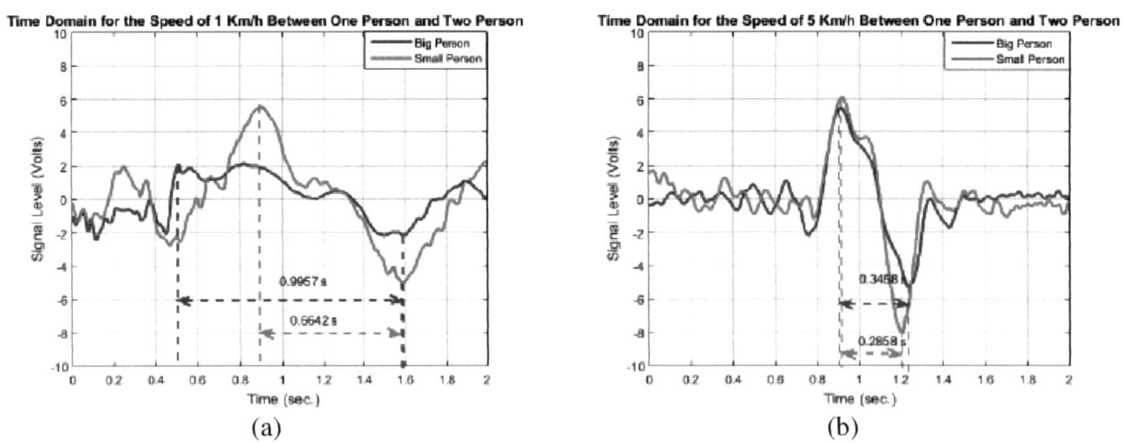

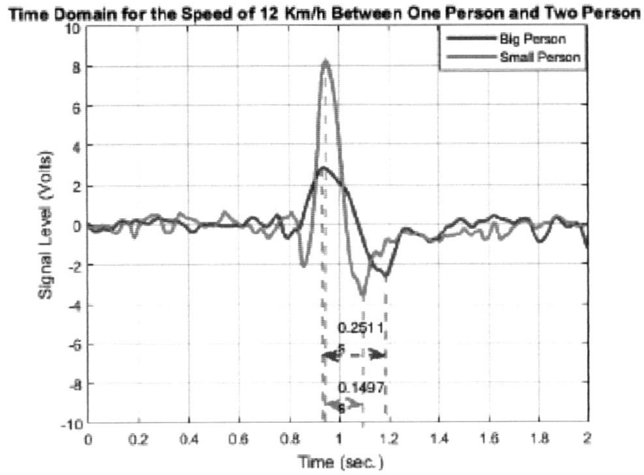

(c)

Figure 7. Time Domain of 1.8 GHz signal at UiTM for the three different speeds of Big Person and Small Person (a) 1 km/h, (b) 5 km/h and (c) 12 km/h.

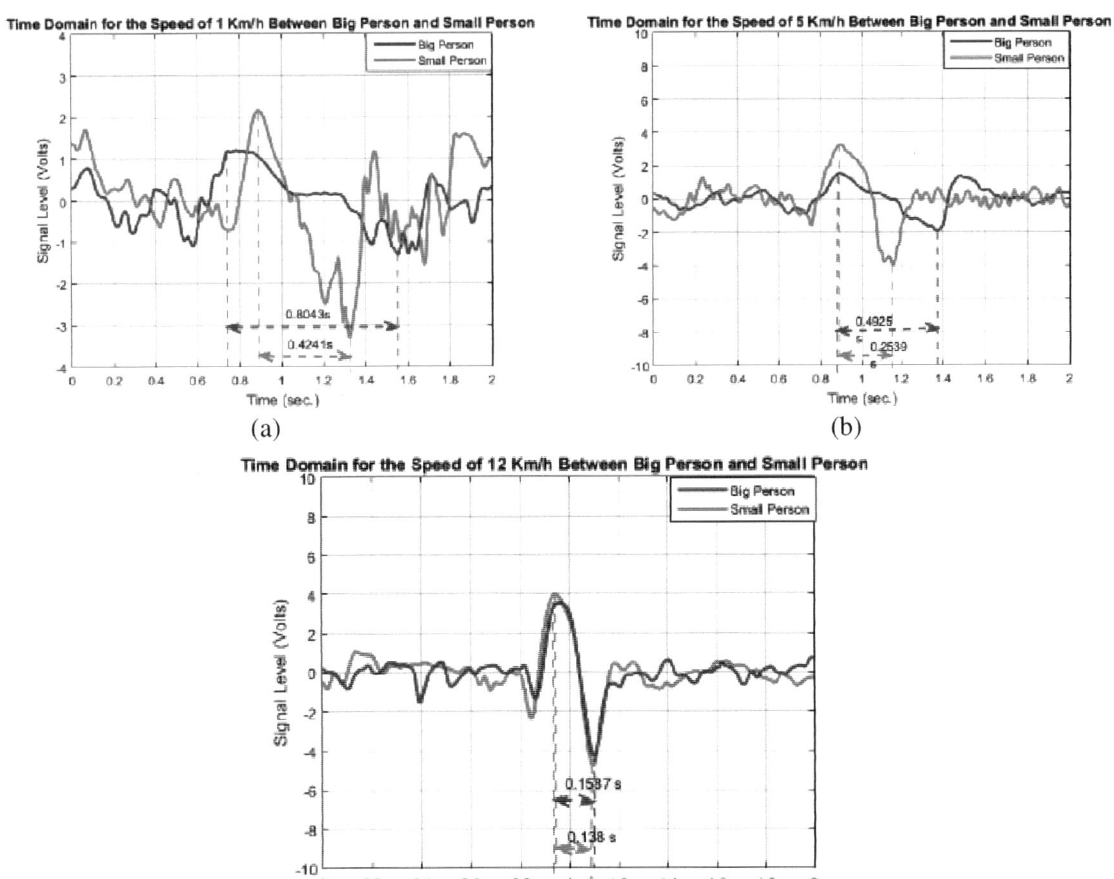

(c)

Figure 8. Time Domain of 1.8 GHz signal at Tanjung Karang for the three different speeds of Big Person and Small Person (a) 1 km/h, (b) 5 km/h and (c) 12 km/h.

3.2. Power Spectral Density (PSD)

The speed and the size of the target can vary the PSD due to the scattering effects of the signal. Figure 9, figure 10 and figure 11 show the PSD results of the three different speeds at three different experimental sites. In term of speed, figure 9 shows that the first lobe of each speed increased as the speed increased. This was proven by the relationship between frequency and period ($f = 1/t$). So, when the time taken for the motorcycle to pass through the line between the transmitter and receiver decreased, the frequency increased. In term of size, since two persons riding a motorcycle had a larger body size, the time taken for the target to finish travelling in the radar range was longer. This resulted the first lobe signal for the two persons (blue line) tended to be narrower compared to the one person (red line) riding the motorcycle.

The PSD results for the next two experimental sites, which are UiTM and Tanjung Karang also followed the trends of the result in MAEPS. Figure 10 and figure 11 show the PSD results of the three different human speeds, which were 1 km/h, 5 km/h and 12 km/h and different body sizes at both of experimental sites. As the speed increased, the first lobe of the PSD grew larger. The higher the target's speed, the shorter the time to finish travelling in the radar range. When the period of time became shorter, the frequency became larger. In term of the person's body size, the bigger the person's body, the narrower the first lobe in PSD compared to the smaller person. The time to travel between the transmitter and receiver for bigger sized human was longer, reducing the frequency compared to smaller sized person.

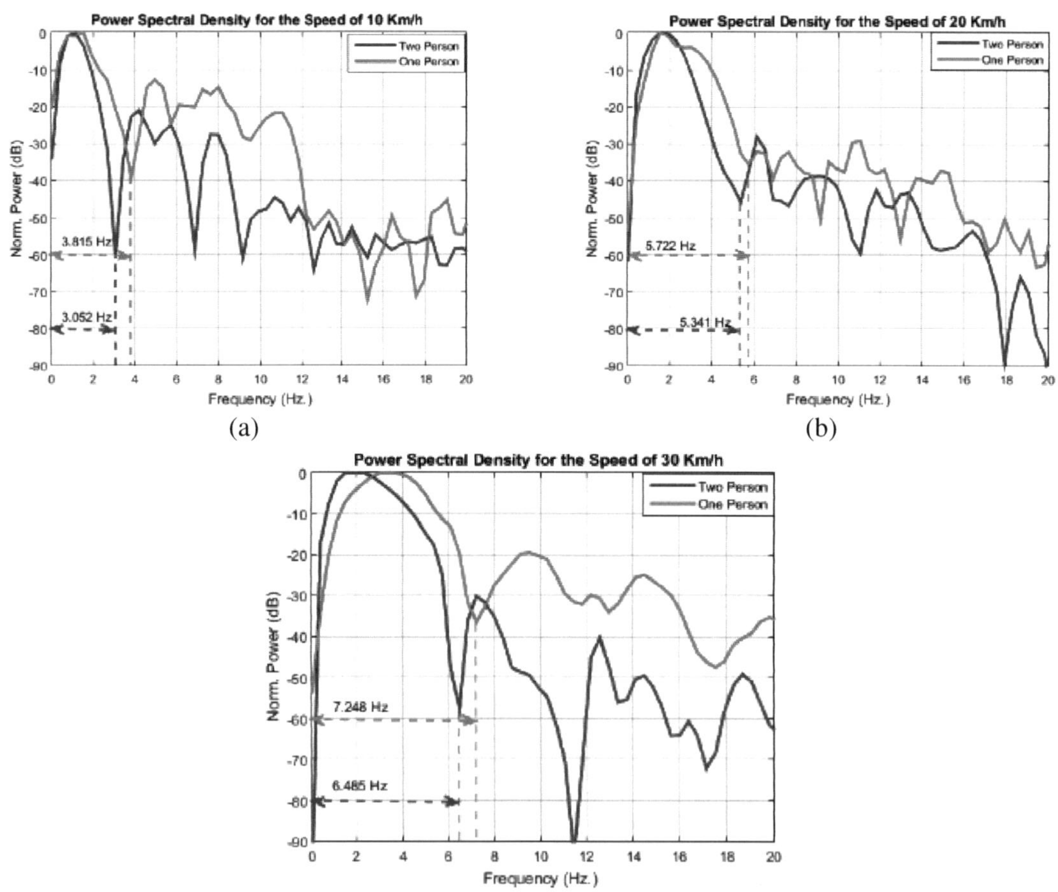

Figure 9. Power Spectral Density of 1.8 GHz signal at MAEPS for the three different speeds of one person and two persons on motorcycle (a) 10 km/h, (b) 20 km/h and (c) 30 km/h.

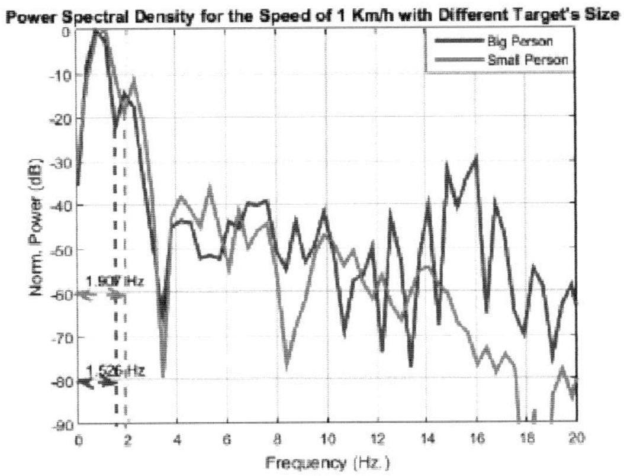

(a)

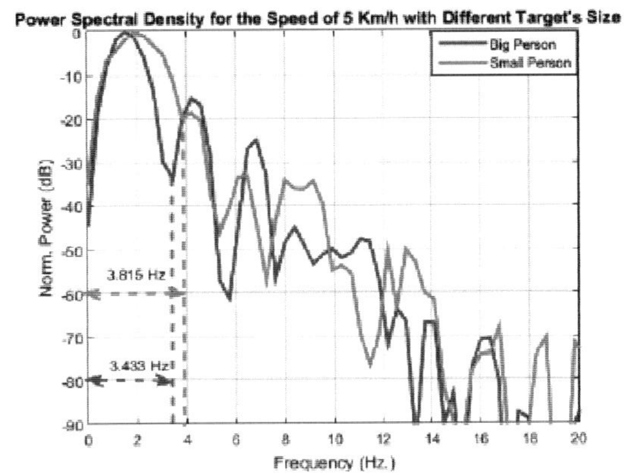

(b)

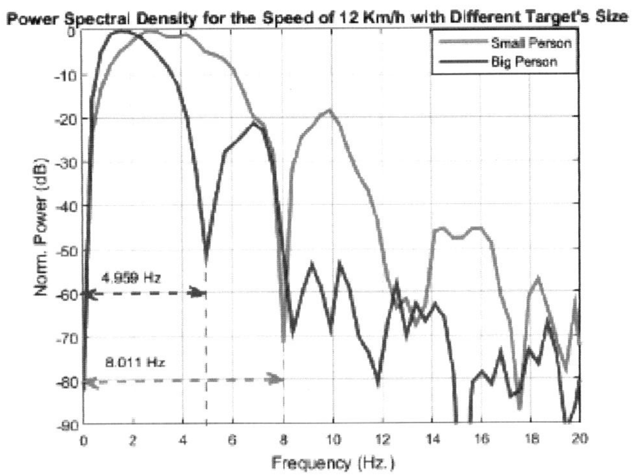

(c)

Figure 10. Power Spectral Density of 1.8 GHz signal at UiTM for the three different speeds of Big Person and Small Person (a) 1 km/h, (b) 5 km/h and (c) 12 km/h.

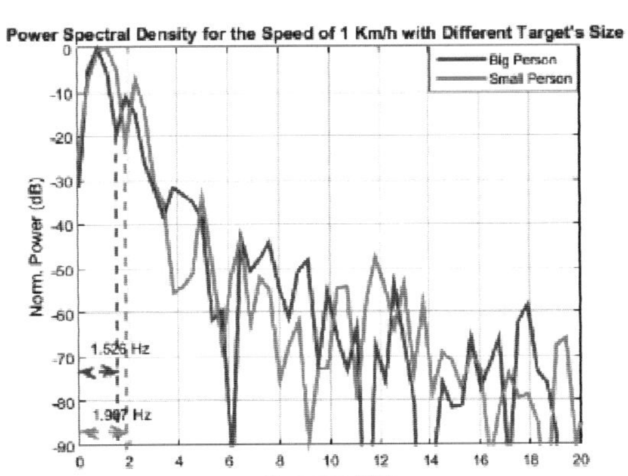

(a)

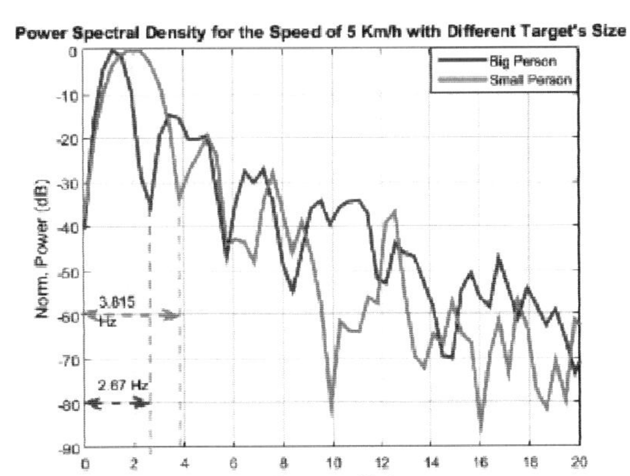

(b)

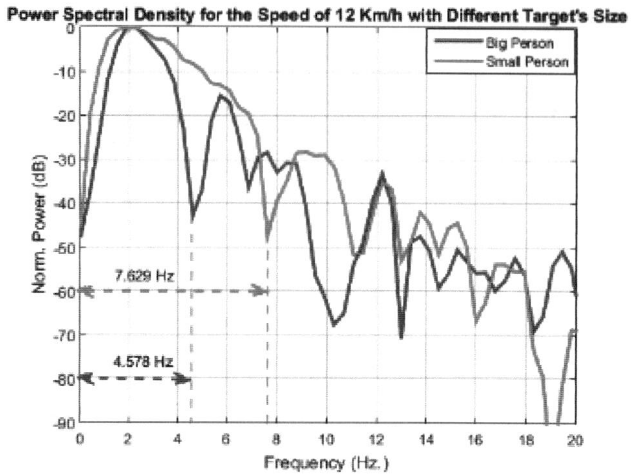

(c)

Figure 11. Power Spectral Density of 1.8 GHz signal at Tanjung Karang for the three different speeds of Big Person and Small Person (a) 1 km/h, (b) 5 km/h and (c) 12 km/h.

3.3. Principal Component Analysis

Principal Component Analysis is a method to classify the data according to its group characteristic and produce a pattern so that the data is easy to visualize [7]. Figure 12, figure 13 and figure 14 show the PCA results for the classification of three different speeds at the three experimental sites. Based on figure 12, the PCA shows a pattern where the 10 km/h (blue) is clustered on the left, the 20 km/h (red) is clustered in the middle and the 30 km/h (black) is clustered in the right for both, one person and two person, riding the motorcycle. This pattern is also the same with the result of PCA for the data collected in Tanjung Karang based on figure 14 displayed. But, for the result of PCA in figure 13, at UiTM, it shows a slightly different pattern for the case in detecting the different speed for the bigger sized person. But the data was still managed to be clustered in a group.

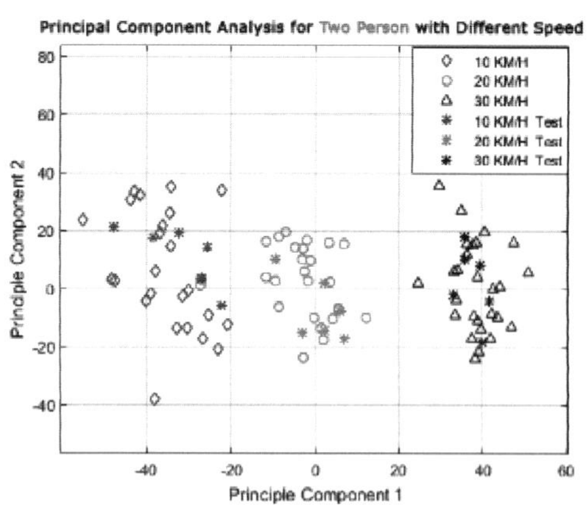

(a)

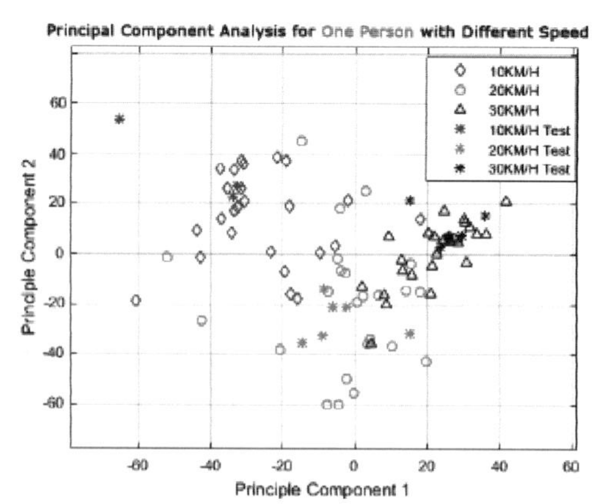

(b)

Figure 12. Principal Component Analysis of 1.8 GHz signal at MAEPS for three different speeds of (a) two person and (b) one person.

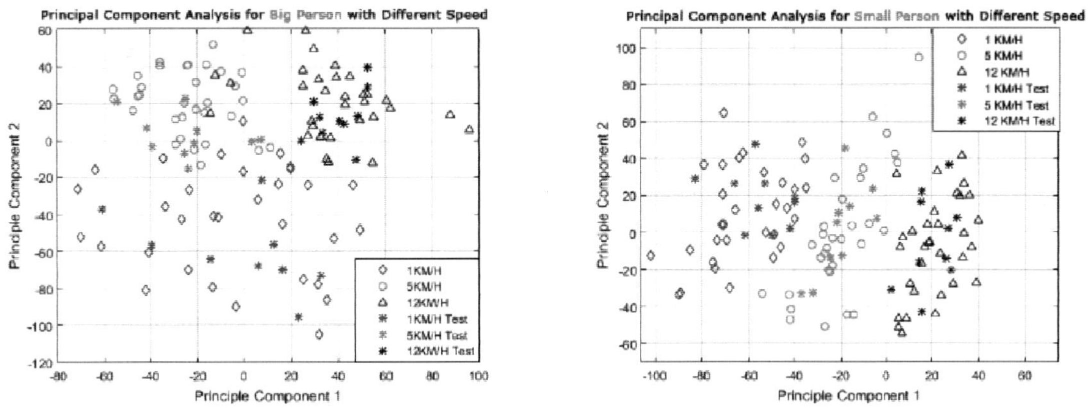

(a) (b)

Figure 13. Principal Component Analysis of 1.8 GHz signal at UiTM for three different speeds of (a) Big Person and (b) Small Person.

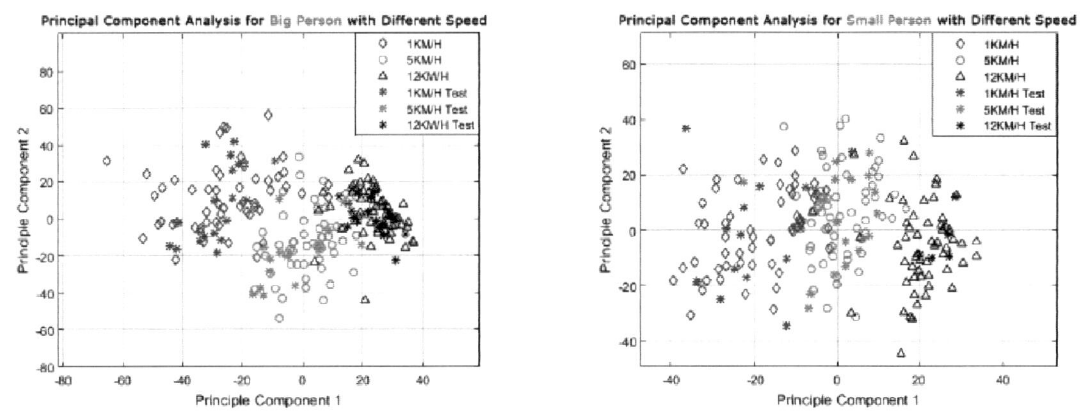

(a) (b)

Figure 14. Principal Component Analysis of 1.8 GHz signal at Tanjung Karang for three different speeds of (a) Big Person and (b) Small Person.

3.4. Confusion Matrix

To study the performance of the classification, the method of confusion matrix was used. The confusion matrix used training data and testing data from PCA to calculate the percentage of performance of the classification. The percentage of performance used ‘three nearest neighbors’ method. Three nearest neighbors work based on the nearest distance of the testing data to the training data. The first three data that were near to the testing data were taken as the part of the percentage performance. Table 3, table 4 and table 5 show the tabulated data of the confusion matrix for each of the three experimental sites.

In table 3 and table 4, the confusion tables display the results of percentage performance of the three different speeds in MAEPS and UiTM. For both body sizes, it achieved 100% percentage performance. This was due to the distance between the target and the transmitter being near (186.77 m and 151.48 m), making strong signal strength to detect the target clearly. The 100% percentage performance indicates that the location and condition was suitable for the radar to operate.

In table 5, the confusion table displays the result of percentage performance of the three different human speeds of the smaller person and bigger person in Tanjung Karang. The percentage performance for this confusion matrix was not all 100% as the previous experimental sites. The worst percentage performance was 86.7% for the classification data of the big person at 5 km/h speed. Based on figure

14, the PCA for the big sized person shows that some of the data for the three speeds overlap each other. This is the reason for the lower percentage performance, because the data did not perfectly cluster into a group. This was due to the distance between the target and the transmitter being far (845.69 m), making the signal strength weak to detect the target clearly. This resulted in a lot of noise in detecting the target Doppler signature.

Table 3. Confusion matrix for 1.8 GHz at MAEPS.

Target's Speed (km/h)	Number of samples				Automatically classified (100%)	
	Training		Testing			
	One Person	Two Person	One Person	Two Person	One Person	Two Person
10	30	33	6	6	100	100
20	33	33	6	6	100	100
30	32	33	5	6	100	100

Table 4. Confusion matrix for 1.8 GHz at UiTM.

Target's Speed (km/h)	Number of samples				Automatically classified (100%)	
	Training		Testing			
	Small Person	Big Person	Small Person	Big Person	Small Person	Big Person
1	42	41	10	9	100	100
5	42	42	10	10	100	100
12	42	42	10	10	100	100

Table 5. Confusion matrix for 1.8 GHz at Tanjung Karang.

Target's Speed (km/h)	Number of samples				Automatically classified (100%)	
	Training		Testing			
	Small Person	Big Person	Small Person	Big Person	Small Person	Big Person
1	73	75	17	19	94.1	94.7
5	75	72	19	16	94.7	86.7
12	61	75	5	19	100	91.7

4. Conclusion

Different target objects have their own Doppler signature. This Doppler signature can be detected and classified into its cluster and form a pattern, so that it can be easily visualized. Based on the results obtained in the study of the speed and the size of the target using passive forward scattering radar, it was proven that the speed and size of the target do indeed affect the characteristic of the Doppler signature. The effect was successfully detected and classified into its own group. Furthermore, the study shows that when the speed of the target was high, the Doppler signature appeared to decrease due to the shorter period of time for the target to pass through the radar range. Shorter period of time produced a higher frequency due to the frequency and time relationship. This system can be further studied so that it can be implemented in real radar detection situation, for example, to detect a very high speed target such as an airplane, high speed boat and even a small size drone that has infiltrated military defense.

5. References

[1] Xu C 2015 *Automatic target detection and speed estimation using forward scatter radar sensor* (University of Birmingham) p 367

[2] Oikonomou D, Nomikos P, Limnaios G and Zikidis K 2019 Passive radars and their use in the modern battlefield *Journal of Computations & Modelling* **9** no. 2 p 37-61

[3] Palmer J E, Harms H A, Searle S J, and Davis L 2012 DVB-T passive radar signal processing *IEEE transactions on Signal Processing* **61** no 8 p 2116-26

[4] Abdul Aziz N H, Raja Abdullah R S A and Muhammad Yusof A N 2017 Human detection and recognition system using passive forward scattering radar *Science International* **29** no 1 p 69–73

[5] Jolliffe I T and Cadima J 2016 Principal component analysis: a review and recent developments *Philosophical Transactions of the Royal Society A: Mathematical, Physical and Engineering Sciences* 374 no 2065 p 20150202

[6] Kataria A and Singh M D 2013 a review of data classification using k-nearest neighbour algorithm *Int. Journal of Emerging Technology and Advanced Engineering* **3** p 354-60

[7] Han K, Hadjipantelis P Z, Wang J L, Kramer M S, Yang S, Martin R M and Müller H G 2018 Functional principal component analysis for identifying multivariate patterns and archetypes of growth, and their association with long-term cognitive development *PloS One* **13** no11 p e0207073

Journal of Physics: Conference Series 2250 (2022) 012009 doi:10.1088/1742-6596/2250/1/012009

Rehabilitation monitoring prototype: Arduino Nano 35 BLE

Mad Kaidi, H.[1]*, Izhar, M.A.M.[1], Ahmad, N.[1], Dziyauddin, R.A.[1], Sarip, S.[1], and Jalil, S.Z.A.[1],
[1]Razak Faculty of Technology and Informatics, Universiti Teknologi Malaysia, Kuala Lumpur, Malaysia

*Corresponding author: hazilah.kl@utm.my, ORCID ID:0000-0001-5915-2875

Abstract. Nowadays, smart health has been developing in the healthcare system by implementing the Internet of Things. One of the implementations of smart health is remote monitoring systems for rehabilitating patients such as stroke. Today, with the rising Covid-19 pandemic, patients undergoing rehabilitation at home have difficulties meeting with their doctors due to the moving restrictions. The healthcare facilities are focused on treating Covid-19 patients. These restrictions have caused doctors and patients not to meet regularly to collect their data on the rehabilitation progress. This research suggests building a prototype to monitor a post-stroke patient's lower limb strength rehabilitation process by using embedded sensors and microcontrollers. The prototype will measure key components of the rehabilitation process and will be discussed in the later section of this paper.

1. Introduction

Industry 4.0 is a representation of the fourth revolution that occurs in manufacturing. This fourth iteration of the industrial revolution will continue the adoption of automation systems and computers that have been implemented in the third revolution and enhance the features by including data and machine learning. In addition, there are a few applications of smart machines introduced by Industry 4.0, one of which is the Internet of Things (IoT) [1].

A system of interlinked computing devices, digital and mechanical machines, objects with unique identifiers, and the capability to transfer data over a network without human-to-human or human-to-computer interaction is the definition of IoT. IoT was introduced in the early 1980s when students from Carnegie Mellon University developed the first internet-connected vending machine thatnotifies users if the drinks are cold enough to be sold [2]. Nowadays, IoT is considered one of the main components of developing smart systems in many organizations and fields.

The IoT ecosystem consists of embedded systems that include processors, sensors, and communication hardware. These components will be used to collect, send, or assess the data they received. These devices work without human interaction as they use smart machine learning or artificial intelligence (AI) but can be accessed by humans if needed [3].

One of the applications of IoT is in the healthcare industry. According to the statistics, more than 60% of medical companies globally have previously acknowledged, and some have adopted IoT in their systems [4-6]. Implementing the IoT in healthcare systems has significantly improved the interaction between patients and doctors. The introduction of IoT has allowed the patients to be continuously monitored by the doctor by receiving live updates of their patients' conditions. The system will make it easier for doctors to make recommendations accordingly.

The organization of this paper will continue in Section 2 as the application of the IoT in the healthcare system. Then, the detailed description and development of Arduino Nano 33 BLE in Section 3. In Section 4, a discussion of the results, analysis, and recommendations for improvement. Then, the paper is concluded in Section 5.

2. Internet of Things in Healthcare

Internet of Things is used for monitoring purposes. This technology helps humans monitor their activities by using small devices and remotely.

2.1. Remote monitoring

Remote monitoring devices in healthcare systems are the most common application of IoT. For example, hospitalized patients who need close attention can be monitored remotely using IoT-driven devices. These devices automatically collect patients' health data such as their heart rate, blood pressure, temperatures, and pain. They will be stored in the clouds to analyze the medical staff. The frequency of the recorded data for each visited patient is reduced as the process is now automated[7]. Besides being at the healthcare facilities, patients undergoing rehabilitation at home can also be monitored using the IoT-driven devices as their data can also be accessed by the doctors remotely. Any changes or disturbances can alert the doctors or the caretakers for immediate response. Thus, it will improve the efficiency and quality of care, reducing the need for the medical staff to visit the patients for data collection, especially during this pandemic.

2.2. Stroke rehabilitation Device

There are a few rehabilitations processes that utilize the smart healthcare system, and one of them is the monitoring system for stroke rehabilitation. This paper suggests building a prototype to monitor the rehabilitation process of a stroke patient, especially for lower limb strength. Patients undergoing the rehabilitation process at home need to conduct exercises to improve their posture and mobility; hence, they need a device to record their progress. Other than that, rehabilitation progress also monitors the recovery of the patients.

This prototype requires the users to interact with it to collect data. Different sensors are attached to the prototype to collect heartbeat and leg strength monitoring data. The components and methods to test the prototype used will be discussed in detail next section.

3. Rehabilitation monitoring prototype using Arduino Nano 33 BLE

The prototype will include microcontrollers as the computing device to process the data collected by the sensors embedded in the systems. These sensors are explicitly chosen to follow the project's requirement to obtain the desired data representing the rehabilitation process of a stroke patient for heartbeat monitoring and less strength at a particular lower limb. The movement of the foot refers to general step movements. However, for details measurement consist of several phases such as heel-strike (HS), foot-flat (FF), midstance (MS), heel-off (HO) and toe-off (TO) [8]. This movement has been implemented in many gait analysis applications involving gait cycles.

The hardware prototype was assembled and programmed to receive input from sensors (pulse and accelerometer). Then the Arduino NANO 33 BLE was booted. Three subjects (Subject A, Subject B, Subject C) were chosen to undergo the testing for data collection. Table 1 tabulates the setup measurement for both procedures.

Table 1. Measurement procedure

	Heartbeat monitoring	Steps counting (accelerometer)
1	The python script (heartbeat_monitoring.py) for the pulse sensor was executed.	The python script (count_step.py) for step counting was executed.
2	The pulse sensor is tied to the thumb of Subject A.	Subject A was walking for 20 steps.
3	The heartbeat of Subject A was manually counted for 60 seconds by placing two fingers between the bone and the tendon over the radial artery.	The step counter from the script was recorded.
4	The manual reading was recorded.	Subject A repeated steps (2) until (3) two more times.
5	The beat-per-minute (BPM) of subject A from the sensor reading was recorded.	Steps (2) until (4) was repeated using 50 steps.
6	Steps (2) until (5) was repeated six times.	The python script was terminated.
7	The python script was terminated.	Steps (1) until (6) are repeated by Subject B and Subject C.
8	Steps (1) until (8) was repeated by Subject B and Subject C.	

The board chosen for the prototype is the Arduino Nano 33 BLE. This board is newly introduced and has many improvements over the regular Arduino Nano. This version of Arduino Nano is redesigned and has the features required to make this prototype. The new central part is the upgraded processor that allows higher data and programs to be stored in the memory. Other than that, the board comes with built-in Bluetooth pairing features via NFC, and it consumes less power making it more efficient to be used in the prototype. Finally, the board has built-in nine-axis internal measurement units such as an accelerometer, a gyroscope, and a magnetometer. Arduino Nano 33 BLE features and small form factor are perfect for this prototype [9].

The finger clip of the heart rate sensor is used in this prototype to measure the heart rate of the stroke patient. Heart rate monitoring for a stroke patient is very crucial. Any disturbance in the rhythm of the heartbeat will be notified to the doctors or caretakers as they will decide if the patient needs immediate action or not. This is a vital part of the prototype, so the data acquired from this sensor must be accurate.

4. Results and analysis

These results are then plotted into graphs for heartbeat monitoring. In the charts as described in Fig. 1, the results obtained from each subject are compared to analyze the accuracy of the sensor. The graph pattern of all three subjects are closely similar between 70-90 bpm.

The next part is to measure the step count of the subjects (off-patient). It is also essential for the stroke patient rehabilitation process. By implementing the built-in accelerometer in the Arduino Nano 33 BLE, a python code is written to track the steps of the subjects. The benchmark for the measures that will be compared is 20 and 50 steps. The subjects then carry the prototype while walking the steps. The result is tabulated in Table 2 and described in Fig. 2.

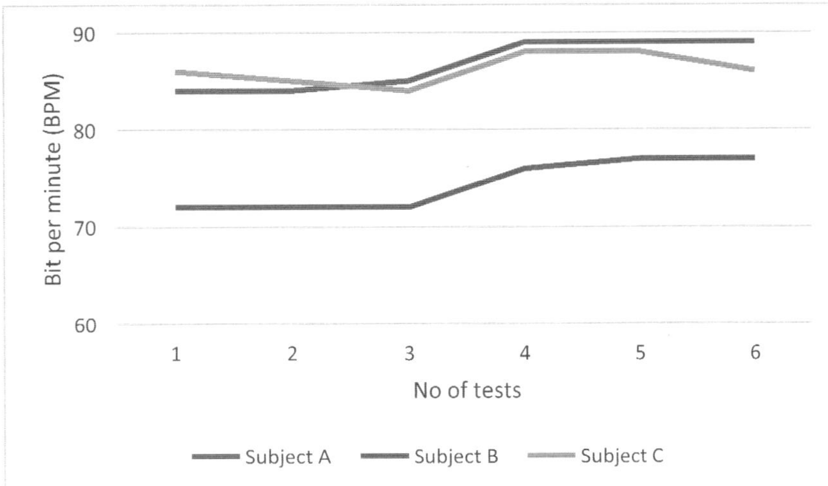

Figure 1. Heartbeat monitoring for various subjects

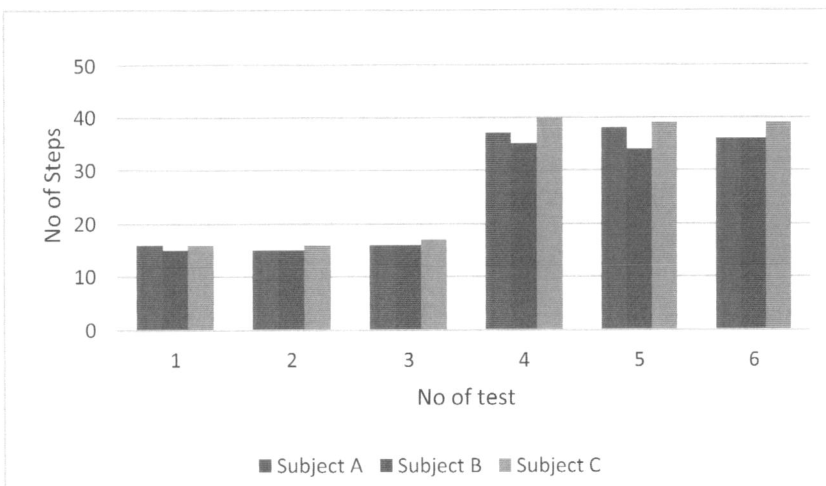

Figure2. Step counting pattern for various subjects

Table 2. Result of step tracking movement

Test #	Real Number of Steps	Accelerometer steps measurement		
		Subject A	Subject B	Subject C
1	20	16	15	16
2	20	15	15	16
3	20	16	16	17
4	50	37	35	40
5	50	38	34	39
6	50	36	36	39

The result shows that the accelerometer is working to register the steps walked by the subjects. The result was consistent throughout the test, but it was not as accurate as of the desired step numbers.

A small error is detected as the subjects walk to 20 steps, but a significant difference in steps detection compared to the 50 steps mark. It might have resulted from a precision fault in the device itself or the program written for the accelerometer. This needs to be revised and improved as the sensor is not yet ready to use in the prototype.

4.1 Recommendation

Based on the obtained result of the experiment conducted using the current prototype, the results are analyzed, and there are a few recommendations made to improve the prototype. The recommendation is to change or add components to the prototype and change the testing methodology. This will be explained in detail in this section.

The lower limb part of the prototype needs to be improved. The result shows that the step counter sensor has produced a less desired output where the steps taken by manual count and the accelerometer are different. In addition, it shows inaccuracy, so this part of the prototype needs to be improved.

Other than that, another sensor can improve the lower limb strength data collection. Here, the force-sensing resistor might be implemented to detect the pressure exerted by the subjects during their walk. It can improve the data analysis of the lower limb strength. If the pressure measurement is high, the subject might progress better in the rehabilitation process. By having this sensor, the pressure imbalance can also be determined to see if the lower limb strength of the subject has imbalances. The force-sensing resistor can be combined into a design to sense the pressure points where the feet touch the floor. Combining these sensors can improve detection, thus providing lower limb strength data accuracy.

5. Conclusion

In conclusion, the lower limb prototype was successfully made, improving from the earlier iteration. It uses less cost and has many improvements from the past model. Based on the result, some parts of the prototype need to be revised, such as adding the pressure sensor and the step tracking sensor, as they did not provide the desired output. Hopefully, this prototype will be ready and available with other sensors such as for upper limb strength, sleep study, etc., to be tested as it will be a good product that can help many people recover from a stroke. This prototype can also inspire others to explore the Internet of Things in the health care system.

References

[1] L. Atzori, A. Iera, and G. Morabito, "The Internet of Things: A survey," Computer Networks, vol. 54, no. 15, pp. 2787–2805, Oct. 2010.

[2] S. Madakam, "Internet of Things: Smart Things," International Journal of Future Computer and Communication, vol. 4, no. 4, pp. 250–253, 2015.

[3] A. Ghosh, D. Chakraborty, and A. Law, "Artificial intelligence in Internet of things," CAAI Transactions on Intelligence Technology, vol. 3, no. 4, pp. 208–218, 2018.

[4] S. Selvaraj and S. Sundaravaradhan, "Challenges and opportunities in IoT healthcare systems: a systematic review," SN Appl. Sci., vol. 2, no. 1, p. 139, Dec. 2019.

[5] H. M. Kaidi, A. H. Abdullah, H. Abdullah, S. Z. A. Jalil, S. Sarip, M. A. M. Izhar, N. A. Bani, "Internet Of Things Devices And Issues In IoT System Development For Healthcare," Turkish Journal of Physiotherapy and Rehabilitation, vol. 32, no. 2, pp. 2836-2843, Mar. 2021.

[6] N. A. Mashudi, H. M. Kaidi*, S. Sarip and L. A. Latiff, "The modelling and simulation of IoT system in healthcare applications," International Journal of Advanced Technology and Engineering Exploration, vol 8, no. 74, 2021.

[7] B. S. Babu, K. Srikanth, T. Ramanjaneyulu, and I. L. Narayana, "IoT for Healthcare," Int. J. Sci. Res., vol. 5, no. 2, pp. 322–326, 2016.

[8] Kharb, A., Saini, V., Jain, Y.K., & Dhiman, S. (2011). A Review Of Gait Cycle And Its Parameters, International Journal of Computational Engineering & Maangement, Vol13, July 2011, 2230-7893.

[9] A. Kurniawan, "Arduino Nano 33 BLE Sense Board Development," in IoT Projects with Arduino Nano 33 BLE Sense: Step-By-Step Projects for Beginners, A. Kurniawan, Ed. Berkeley, CA: Apress, 2021, pp. 21–74.

Acknowledgement

Universiti Teknologi Malaysia supports this work under research University Grant Q.K130000.3556.06G45 for the financial support provided throughout this research project.

Fabrication and characterisation of Frequency Selective Surface (FSS) for Terahertz Sensing Application

N S Ishak[1], N Zainal[1,2*], F C Seman[1,3] and S M Shah[1,3]

[1] Faculty of Electrical and Electronics Engineering, Universiti Tun Hussein Onn Malaysia, Parit Raja, Batu Pahat, 86400, Malaysia.
[2] Advanced Sensing Device and Technology (ADST), Universiti Tun Hussein Onn, Parit Raja, Batu Pahat, 86400, Malaysia.
[3] Research Center for Applied Electromagnetics (EMcenter), The Institute of Integrated Engineering, Universiti Tun Hussein Onn Malaysia, Parit Raja, Batu Pahat, 86400, Malaysia.

*Corresponding author: nurfarina@uthm.edu.my

Abstract. A frequency selective surface (FSS) structure based on a circular slot with 28 to 31 μm was fabricated using Electron Beam Lithography combined with wet etching technique on a 400 μm thick quartz substrate. The circular slot pattern, designed using the CST software, was transferred during the Electron Beam (EBL) exposure with 100 kV voltage acceleration, 2 nA of e-beam current, and line dosage of 1000 $\mu C/cm^2$. A 0.5 μm thick aluminium layer was deposited on a quartz substrate using the thermal evaporator technique. The aluminium thickness was confirmed by laser microscope and surface profiler measurement. The investigation shows that 20 seconds is the optimal etching time for producing the desired FSS circular slot structure compared to 40 and 60 seconds. The fabricated samples surface metrology was examined using an optical microscope, Field Emission Scanning Electron Microscope (FESEM) and Atomic Force Microscope (AFM). The material compositions of the samples were confirmed by using the Energy Dispersive X-ray (EDX) measurement. Results show that, on average, there are ±1.5 μm of tolerance produced for fabricated sample size on the circular slot pattern compared to the CST simulation results. FSS performance slightly shifted in the frequency from actual 1.80 THz to 1.79 THz, whereas the transmission magnitude has increased by 0.03 from 0.91 to 0.94.

1. Introduction

Frequency Selective Surface (FSS) is made up of a periodic element consisting of either patch or slot elements. It is widely known for its applications in electromagnetic (EM) shielding [1], microwave filters [2]–[4], absorbers [5], radomes [6]–[8] and radar cross-section [9]. FSS is introduced and applied as a transducer in the biochemical area, which allows for better detection of fungi, bacteria, and pesticide residue [10], [11]. According to the researchers [11]–[13], applying the FSS structure helps increase the detector sensitivity, which lies 10^4 to 10^{10} times higher than the detector without the FSS structure. In communications, the FSS structure was commonly used in the microwave frequencies application, and it has a large structure. However, for higher frequencies, such as in the Terahertz (THz) application, a small FSS structure is required. This can be produced by using the microfabrication process and technologies. Thus, with the possibilities of higher sensitivity in sensing the miniature features and producing the FSS structure in a smaller size, it has been demanding components in THz applications.

There are many ways to fabricate the FSS structure, such as photolithography, lithography, and laser technology. However, the photolithography process is the most common process used in the fabrication of the FSS. There are multiple steps required for photolithography, and one of the steps in the preparation of the mask may affect the size structures of the FSS. For example, these multiple steps may cause a structure pattern error even due to the slight misalignment during the masking process. Hence, laser technology without masking can be an alternative to avoid misalignment errors but requires higher costs [14].

On the other hand, the lithography process can replace photolithography since no mask process is needed. The difference between lithography and photolithography process is lithography does not require any shadowing mask (mask-less) to transfer the pattern structure onto the substrate surfaces. Meanwhile, the photolithography process requires the mask to transfer the structure pattern onto the substrate [15].

The majority of FSS structures with resonant frequency below 1 THz are produced using microfabrication and technology, namely photolithography and wet etching techniques. For instance, researchers [16]–[18] have used photolithography and wet etching techniques to fabricate their FSS structure at a resonant frequency with the range of 0.42 THz, 0.5 THz, and 0.85 THz. Meanwhile, in the case of lithographies such as the nanoimprint and Electron Beam Lithography (EBL), only a small number of researchers have used these techniques [19]. In this research, the lithography and wet etching technique were selected to pattern the micro-size FSS circular slot structure at resonant frequency above 1 THz. The circular slot pattern was chosen due to its less complex structure. At the same time, the EBL technique can minimize and avoid the misalignment issue in transferring the micro-size pattern onto the substrate. During the fabrication process, the wet etching technique was utilized to remove the unwanted metal layer of the FSS pattern. The micro-size pattern of FSS structures will be varied to investigate the fabrication tolerance compared to the simulation results, hence aiming it to be used as a potential future terahertz sensing application.

2. Methodology and experimental details

This study uses a 3-inch quartz wafer, 400 μm-thick with Z-cut mono-crystalline single-sided polished wafer. The quartz sample was then cut into square substrates with 1 cm x 1 cm cut using a diamond cutter. Before the fabrication process, the samples were cleaned by a standard cleaning procedure. First, the square substrates were immersed into acetone, ethanol and isopropanol liquid solution and rinsed with deionized water. Next, the samples were dipped into the hydrofluoric acid (HF) solution for native oxide removal. The samples were rinsed with deionized water, blow-dried with nitrogen gas, and dried in the oven at 90 °C for 5 minutes. A 0.5 μm thick aluminium (metal) layer was then deposited using ULVAC VPC-1100 Thermal Evaporator System. The metal evaporation parameters were set up with a melting point at 660 °C, density of 270 g/cm^3, and Z-ratio 1.08 at the deposition controller monitor. The KLA Tencor D-100 Surface profiler was used to measure and confirm the metal thickness.

Polymethyl Methacrylate (PMMA) was used as a resist layer and spin-coated onto the sample's substrate. The spin coater system was set up with consecutively three cycles, starting with it ramping up to 1000 rpm for 10 seconds, then increasing it to 4000 rpm for 60 seconds, and lastly ramping down to 1000 rpm for 10 seconds. The samples were then baked in the oven for 30 minutes at 90 °C temperature for hardening the adhesive layer onto the substrate surface. The coated samples were exposed with an electron beam using the lithography (EBL) technique. The circular slot pattern designed using the CST software was transferred during the Electron Beam (EBL) exposure with 100 kV voltage acceleration, 2 nA of e-beam current, and line dosage of 1000 μC/cm^2. The resist layer thickness was checked using a surface profiler machine. For pattern development process, the post-exposed samples were then immersed into the mixed developer solution of Methyl Isobutyl Ketone (MIBK) and isopropanol with a ratio of 1:3 for 60 seconds. Next, the samples were rinsed with an isopropanol solution. The removal of the unwanted metal layer, JT Baker aluminium etchant consists of the mixture 71.9 % phosphoric acid, 2.7 % nitric acid, 2.8 % acetic acid, and deionized water was used [20]. The immersing time in the aluminium etchant was varied in obtaining the optimal metal etching time accordingly to the desired circular slot pattern. Acetone was used for removal of the PMMA resist layer, and subsequently, the samples were rinsed with isopropanol, deionized water, and dried up with nitrogen blow. The optical microscope and Field Emission Scanning Electron Microscope (FESEM) were used to check and inspect the samples. Figure 1 illustrates the FSS samples fabrication process.

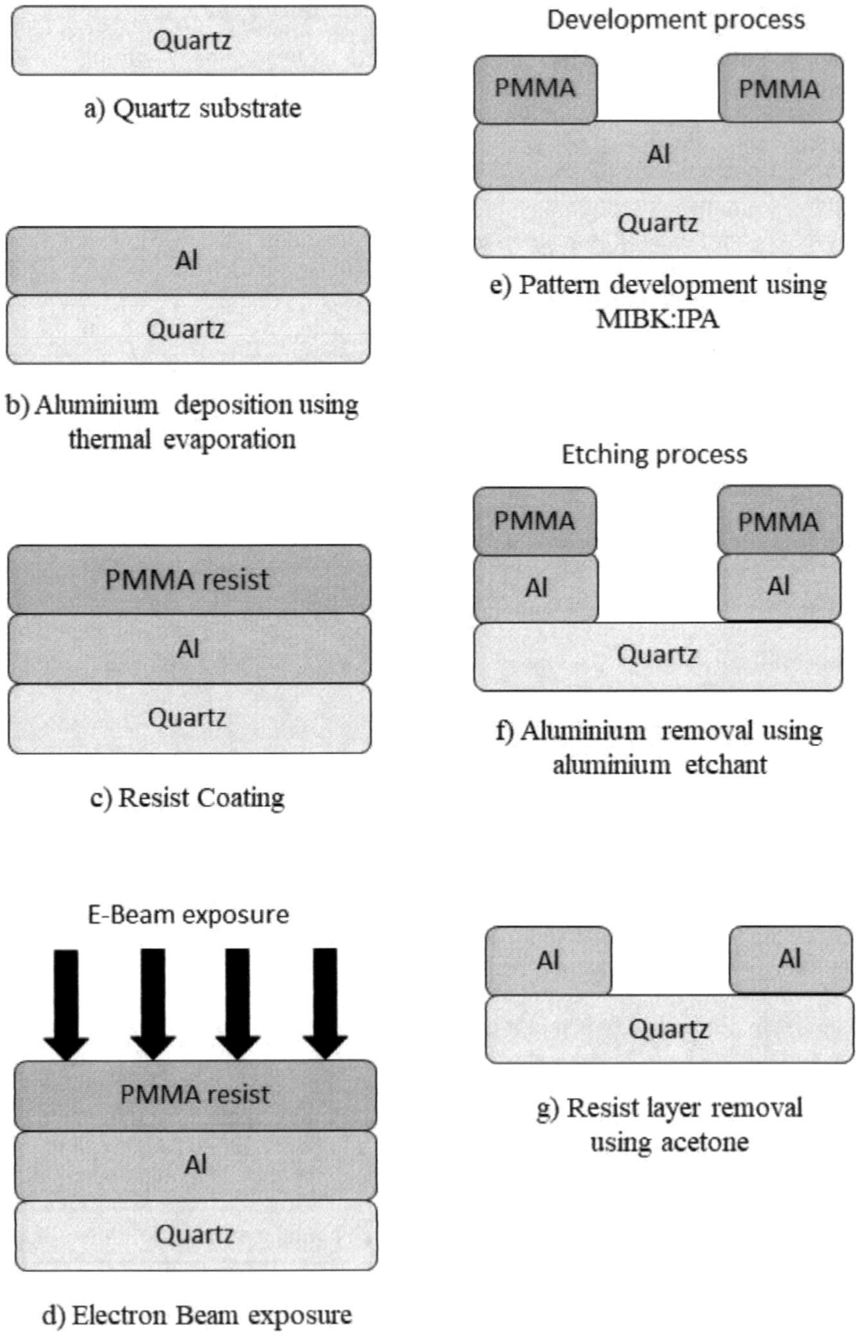

Figure 1(a-g). The overall fabrication process flow of micro-size FSS circular slot structures.

3. Result and Discussions

The aluminium deposition process has been set up and monitored using the ULVAC VPC-1100 thermal evaporator system via deposition controller. Results show that approximately 0.50 µm thick aluminium layer has been deposited onto the quartz substrate for a 1-hour evaporation process. The samples thickness was measured and confirmed using a surface profiler and a laser microscope. Figure 2 shows the five mapping points on the sample surface during the thickness measurements.

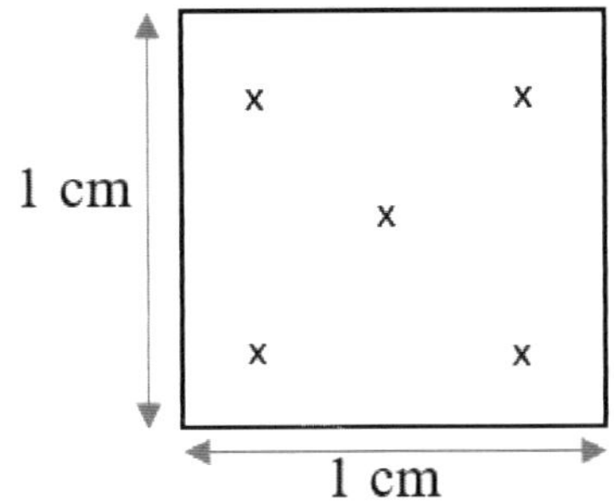

Figure 2. Mapping point during the thickness measurement.

In this study, three different dimensions radiuses of the micro-size FSS circular slot structures were designed using the CST software. The FSS circular slot radius parameter is selectively designed as the size structure will affect its frequency application. In theory, the standard frequency formula can be calculated as $f = \frac{c}{\lambda}$, where, c is the speed of light, and λ is a wavelength. However, in this study, the frequency of the FSS circular slot structure, f_r is calculated using the equation (1) [21] by taking into consideration the speed of light, c, the circumference of the FSS circular slot structure, $2\pi r$, as well as the dielectric permittivity, ε_r. According to [21], the total length of the full complete circular structure can be equal to one wavelength, λ. Figure 3 shows the image of a total length full complete circular slot, which can be referred to as a circumference expressed in equation (2) [22], where, r, is a radius of a circle, and d, is a diameter of a circle. Thus, based on the equation 1, three different radiuses, r of the FSS circular slot have been designed namely patterns 1, 2 and 3 that having 28.27 μm, 38.04 μm, and 31.10 μm as depicted in figures 4 (a), 4 (b) and 4 (c), respectively.

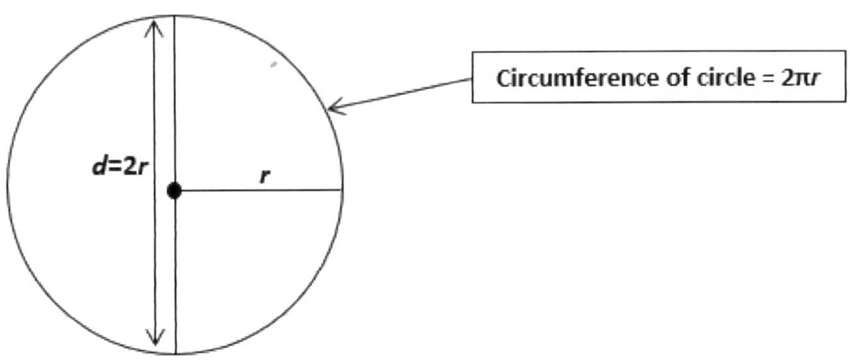

Figure 3. Image of circumference of a circle.

$$f_r = \frac{c}{\lambda\sqrt{\varepsilon_r}} = \frac{c}{2\pi r\sqrt{\varepsilon_r}} \quad (1)$$

$$\text{Circumference of cirlce}=2\pi \text{r, where ;} \quad d = 2r \quad (2)$$

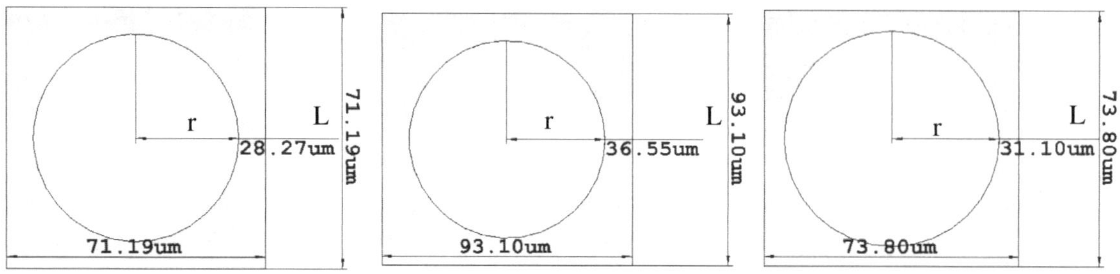

(a) L = W= 71.19 µm, r = 28.27µm

(b) L = W= 93.10 µm, r = 36.55 µm

(c) L = W= 73.80 µm, r = 31.10 µm

Figure 4. The design pattern of micro-size FSS slot structure with size (a) 28.27 x 71.19 x 71.19 µm, (b) 38.04 x 93.10 x 93.10 µm, and (c) 31.10 x 73.80 x 73.80 µm.

The FSS circular slot pattern variation is to investigate the fabrication tolerance for the THz frequency applications. All of the samples have been observed under the optical microscope before and after the etching process. This is to investigate the effect of the etching process on the size pattern. In this study, a wet etching technique was used to develop the FSS slot circular pattern. The wet etching technique was expected to produce the isotropic pattern [23]–[25]. It has an undercut compared to the anisotropic pattern, as shown in figure 5.

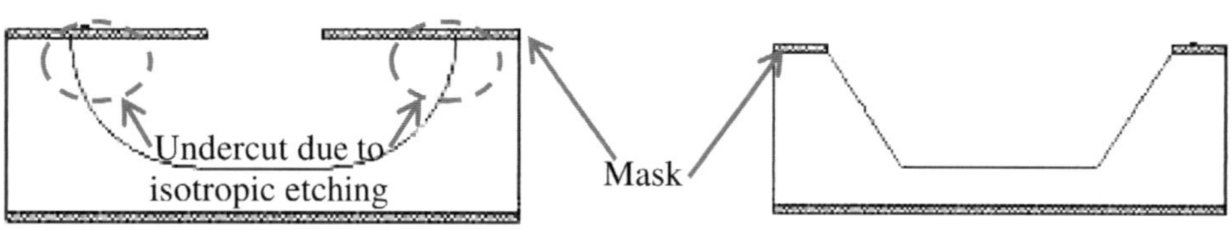

(a) (b)

Figure 5. Difference between (a) isotropic and (b) anisotropic etching.

In the beginning, the metal etching time was set up to 60 seconds for the metal removal. The 60 seconds etching time parameter was commenced by referring to article by [26] has used 5 minutes for metal removal with the sample size of 2 cm x 2 cm and design structure of approximately 110 µm. However, it can be observed from the samples that there is an over-etched pattern examined with the optical microscope for 60 seconds etching time, as shown in figure 6 (c). The etching time was then varied to 40 seconds and 20 seconds. From the investigation, 40 seconds of etching time still produced an over-etched pattern. The circular slot array pattern has shown a combination of each other, as shown in figure 6 (b). The over-etched was expected due to the isotropic etching that has high tendencies to produce undercutting of the resist layer during the process if the sample was immersed into the etching solution for a longer time.

On the other hand, 20 seconds of etching time produced an excellent circular slot pattern as desired design, inspected using the optical microscope. Good circular slots were obtained for all three different patterns and sizes. Thus, these results show 20 seconds it as an optimal etching time for this study. Figure 6 (a) shows the image of good circular slot patterns with 20 seconds etching time, and examined via optical microscope.

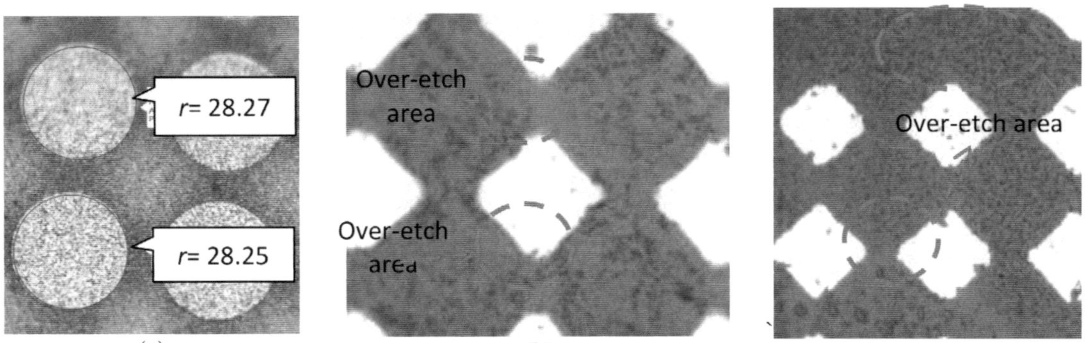

(a) (b) (c)

Figure 6. The images of FSS structure under an optical microscope (a) 20 second, (b) 40 second, and (c) 60 second.

Samples were also inspected using Field Emission Scanning Electron Microscope (FESEM) to observe and measure the actual dimension circular slot after aluminium and resist removal. Figure 7 shows the result from the FESEM image, where the circular slot appears like a patch. The Energy Dispersive X-ray Analysis or EDX measurement was used to identify the samples' element composition, namely the aluminium and quartz substrate. During the EDX examination, a few selected areas were mapped on the samples labelled as points 1, 2, 3 and 4. Point 2 was mapped onto the area covered with aluminium layer, thus representing the reference point in detecting the aluminium composition. Points 3 and 4 were mapped at the area between circular slots at different positions. Hence it is expected to trace the aluminium composition. Point 1 was mapped on the circular slot, where a quartz trace was anticipated to be detected. Figure 8 (b) shows the result EDX image with all mapping points. For better understanding, the samples' cross section and top view EDX images were also provided and shown in Figures 8 (a) and (b), respectively. Results of the EDX material compositions were achieved and detected as expected.

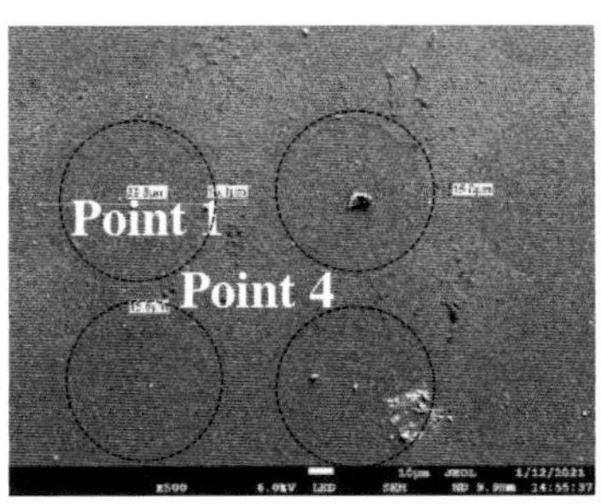

Figure 7. Image of FSS structure observed by FESEM after Aluminium etches and resist removal.

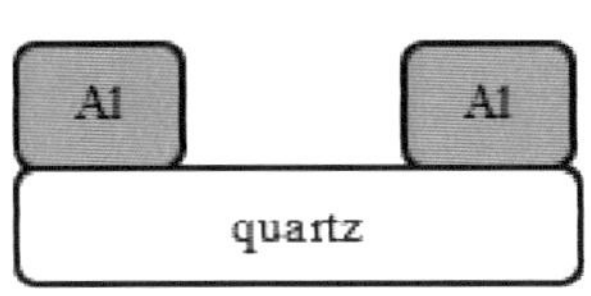

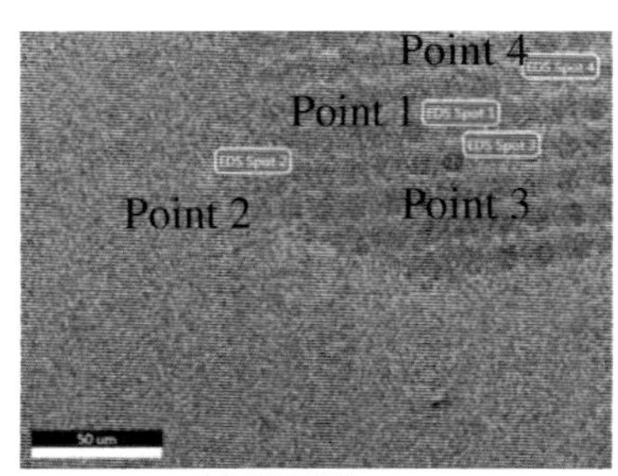

(a) Cross-section view (b) EDX image

Figure 8. FSS sample under the inspection of EDX analysis image.

Investigation under an Atomic Force Microscope (AFM) was also performed, this is to ensure that all the aluminium at the slot area was removed. The result shows that all unwanted aluminium layer has been removed at the slot area, as shown in figure 8. The result in figure 9 can be confirmed by looking at the thickness (nm) according to the colour in the result based on the selective area measurement.

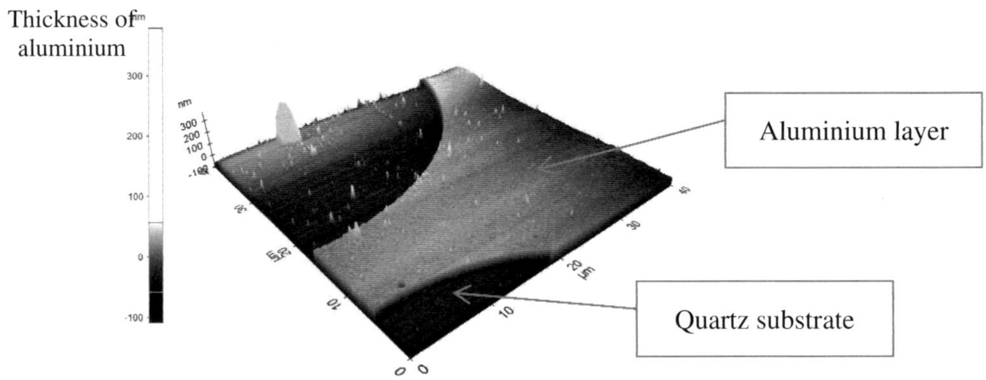

Figure 9. FSS sample under the inspection of AFM.

Due to the undercut produced from the wet etching process, the fabrication tolerance was calculated by comparing the circular slot radius between the actual size pattern (the CST design) and fabricated samples. Table 1 shows the comparison radius measurement before and after the etching process. It was observed that the wet etching technique had affected the radius pattern by having a slightly bigger size, around ±1.5 μm of tolerance.

Table 1. The measurement of FSS radius circular slot between the actual size of CST design (before etching) in comparison to fabricated sample (after etching).

Parameter	The radius of the circular slot in CST (μm)	The radius of the circular slot after etching (μm)	Fabrication tolerance
Pattern 1	28.27	29.76	±1.49
Pattern 2	36.55	38.04	±1.49
Pattern 3	31.10	32.61	±1.51

To investigate the effect of fabrication sample size for the FSS performance, all the fabrication tolerance was reinserted into CST simulation, and the performance was remeasured. The simulation results show slight shifts in the frequency and the transmission magnitude due to the ±1.5 μm of fabrication tolerance size. It is important to compare the fabrication tolerance dimension with the CST design because the FSS will be implemented for future material sensing applications at the Terahertz range. For example, circular slot pattern 1 shows shifts slightly in the frequency from 1.80 THz to a lower range of 1.79 THz, as shown in figure 10. In the case of transmission magnitude, results show an increase of 0.03 magnitude, which increases from 0.91 at 1.80 THz to 0.94 at 1.79 THz. Pattern 2 and 3 circular slots also have similar results as pattern 1.

Figure 10. Comparison between actual and with fabrication Tolerance of circular slot FSS sample

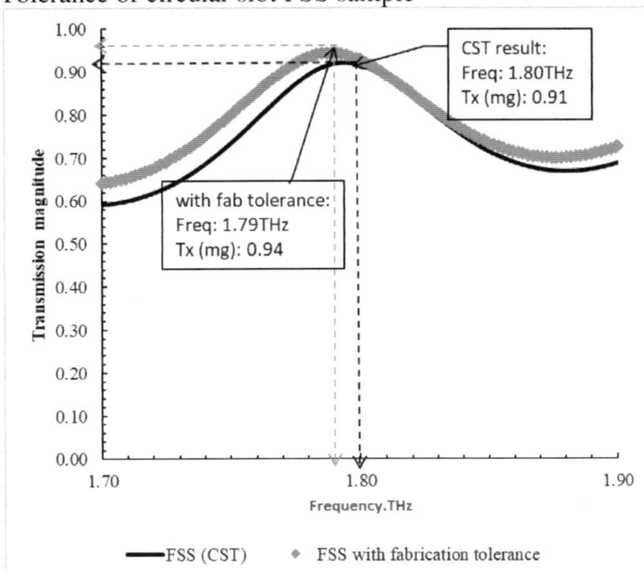

4. Conclusions and Recommendation

In conclusions, this study has successfully fabricated a FSS circular slot structure using Electron Beam Lithography combined with the wet etching technique for THz application. The results show that an approximately ±1.5 µm size of fabricated tolerance was obtained which is due to the isotropic etching. This fabrication tolerance has slightly shifted the frequencies and transmission magnitude in the FSS performance that applied in THz range. Further investigation using the dry etching technique can be carried out in minimising the fabricated tolerance as it produce anisotropic etching.

5. References

[1] Narayan S, Nair R U, and Jha R M 2009 A novel EM analysis of free-standing rectangular aperture FSS based on MM-GSM technique *Applied. Electromagnetic. Conf.* **3** 1–3

[2] Wu T K Improved broadband bandpass FSS filters for 5G applications 2018 *IEEE Antennas Propag. Soc. Int. Symp. Usn. Natl. Radio Sci. Meet. APSURSI 2018 - Proc.* 2033–2034

[3] Chakrabarti T Sarkar S Ghosh S K and Bhattacharyya S An ultra-thin FSS bandpass filter in terahertz region 2019 *TEQIP - III Spons. Int. Conf. Microw. Integr. Circuits, Photonics Wirel. Networks, IMICPW* 492–494

[4] Sun C and Li J 2011 Bandpass filter based on metamaterials in terahertz wave region *4Th IEEE Int. Symp. On Microwave, Antenna, Propagation And EMC Technologies For Wireless Communications*

[5] Mishra R Sahu A and Panwar R Cascaded graphene frequency selective surface integrated tunable broadband terahertz metamaterial absorber 2019 *IEEE Photonics J.* **11** 1-10

[6] Liu N, Sheng X, Zhang C and Guo D 2018 Design of frequency selective surface structure with high angular stability for radome application *IEEE Antennas Wirel. Propag. Lett.* **17** 138–141

[7] Li T W, Li D, and Li E P 2017 A novel FSS structure with high selectivity and excellent angular stability for 5G communication radome *10th Glob. Symp. Millimeter-Waves, GSMM 2017* 50–52

[8] Zhou H et al. 2012 Filter-antenna consisting of conical FSS radome and monopole antenna *IEEE Trans. Antennas Propagation.* **60** 3040–3045

[9] Xue J, Jiang W and Gong S 2017 Wideband RCS reduction of microstrip antenna based on 2.5-dimension miniaturized frequency selective surface *5th Asia-Pacific Conf. Antennas Propagation, Conf. Proc. (APCAP)* 209–210

[10] Qin B et al. 2018 Highly sensitive detection of carbendazim by using terahertz time-domain spectroscopy combined with metamaterial *IEEE Trans. Terahertz Science and Technology.* **8** 149–154
[11] Qin J, Xie L, and Ying Y 2016 A high-sensitivity terahertz spectroscopy technology for tetracycline hydrochloride detection using metamaterials *Food Chem.* **211** 300–305
[12] Park S J et al. 2015 Detection of microorganisms using terahertz metamaterials *Sci. Rep.* **4** 4988
[13] Xie L et al., 2015 Extraordinary sensitivity enhancement by metasurfaces in terahertz detection of antibiotics *Sci. Rep.*, **5** 1–4.
[14] Lin Y et al., 2017 Free-standing double-layer terahertz band-pass filters fabricated by femtosecond laser micro-machining *Opt. Express*, **25** 25125-25134
[15] Chen X et al 2012 Photothermal direct writing of metallic microstructure for frequency selective surface at terahertz frequencies *Proc. 2012 Int. Work. Metamaterials, Meta* 1-3
[16] Zhu Y et al., 2011 THz time-domain spectroscopy of multilayer filters *Int. Conf. on Infrared, Millimeter and Terahertz Waves* 1-2
[17] Wang D S, Chen B J and Chan C H 2016 High-Selectivity Bandpass Frequency-Selective Surface in Terahertz Band *IEEE Trans. on Terahertz Science and Technology* **6** 284-291
[18] Ebrahimi A et al., 2015 Second-Order Terahertz Bandpass Frequency Selective Surface With Miniaturized Elements *IEEE Trans. on Terahertz Science and Technology* **5** 761-769
[19] Betancourt T and Brannon-Peppas L 2006 Micro- and nanofabrication methods in nanotechnological medical and pharmaceutical devices *Int J. of nanomedicine* **1** 483–495
[20] Imaeda K, Bessho K and Shikida M 2016 Design and fabrication of differently shaped pyramids on Si{100} by anisotropic wet etching *Microsystem Technologies* **22** 2801–2809
[21] Seman F C and Khalid N K 2014 Design strategy for optimum planar square loop FSS with different dielectric substrates *Theory and Applications of Applied Electromagnetics* (Springer: International Publishing)
[22] Burzynski D and Ellis W 2008 *Fundamentals of Mathematics* OpenStax CNX
[23] Pal P and Sato K 2010 Fabrication methods based on wet etching process for the realization of silicon MEMS structures with new shapes *Microsystem. Technologies* **16** 1165–1174
[24] Negi S, Bhandari R and Solzbacher F 2011 Novel isotropic and anisotropic etching of MEMS structures by controlling the dynamics of the etchant *16th Int Solid-State Sensors, Actuators and Microsystems Conf*, 2362-2365
[25] May G S and Sze S, 2004 *Fundamentals of Semiconductor Fabrication* (Wiley, John Wiley & Sons, Inc).
[26] Ferraro A et al, 2017 Broad- and Narrow-Line Terahertz Filtering in Frequency-Selective Surfaces Patterned on Thin Low-Loss Polymer Substrates *IEEE J. Sel. Top. Quantum Electron*, **23**.

Acknowledgments
The authors wish to acknowledge the funding of this project by Fundamental Research Grant (FRGS/1/2017/TK04/UTHM/02/13) and GPPS Grant VOT H026, Ministry of Education Malaysia, Registrar Office, and Research Management Centre (RMC) of Universiti Tun Hussein Onn Malaysia (UTHM).

Novel module architecture for wideband multichannel multi band down conversion with built in Local oscillators

Madiwalesh Pattar[1,2], **Gaurav Anand**[1,3]
[1]Product Development and innovation centre, Bharat Electronics Limited, Bangalore, Karnataka-560013, India
[2]madiwaleshmp@bel.co.in, [3]gauravanand@bel.co.in

Abstract. Wide band down converters employed in the field of electronic warfare (EW) process multiple channels simultaneously and in each channel RF signal is split and processed in multi bands. In each channel, RF signal ranging from L to C Band is split into three bands as L, S & C Bands. L Band is directly processed S & C Bands are down converted to identical Intermediate frequency (IF) in UHF to L Band using two different local oscillators (LOs) generated internally. Local Oscillator with low Phase Noise did not introduce jitter on RF input. Module architecture described here is for four channels, extendable to multiple channels.

1. Introduction

Electronic Warfare (EW) RF sub systems are multi-channel and wideband in nature. Four antennas are used in Direction finding (DF) systems to cover entire azimuth, hence most of systems employ quad channel down conversion systems. Airborne EW sub systems must be compact, modular, power efficient and repeatable in performance. Wideband is the basic nature of EW system to identify various sources of foe RADAR transmissions, with better sensitivity and selectivity; each channel wide band RF input is separated and processed as multi-bands. Down converters are always associated with LO generation & associated distribution sections. LO levels are generally higher and need to be routed to each channel. RF cabling for LOs routing will add complexity and make the system bulky. High power level of LOs may radiate in the EW system suite and create EMI/ EMC compatibility issues. Hence it becomes essential to incorporate LOs into the down conversion system itself. LO unit integration poses several challenges to down conversion architecture in terms of effective size utilization, co routing multiple RF & LO bands/ Digital/ Power supply on multiple composite multilayer Printed Circuit Boards (PCB) and interface with each other, isolation among various types of signals.

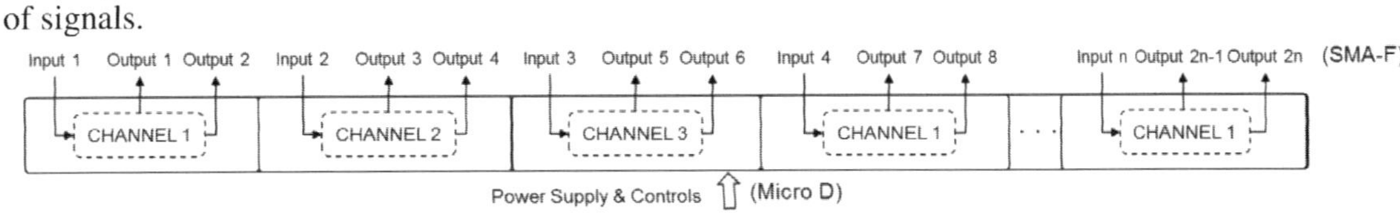

Figure 1: Input Output types and details.

Input-output details of proposed architecture are as shown in Figure 1. RF input in each channel is from L to C Band. It is separated into three bands. Spilt band 1 as L Band using switch, spilt band 2 as S Band & spilt band 3 as C Band using diplexer. In each chain there is one amplification & filtering chain for L Band and two RF conversion chains for S & C Bands. RF signal is amplified suitably, filtered and given to mixer for down conversion with respective LOs. LO1 & LO2 down converts spilt band 2 & 3 respectively to identical IF in UHF to L Band. [2] Discusses RF Band wise simulation, component level implementation, test results and its compliance against specifications.

Journal of Physics: Conference Series 2250 (2022) 012011 doi:10.1088/1742-6596/2250/1/012011

Table 1. Specifications

Parameter	Specification
RF Input Frequency range, No of channels	L to C Band, 4 Channels, Four Inputs (SMA-F)
Output intermediate Frequency (IF) range	UHF to L Band, Eight Outputs (SMA-F)
Instantaneous IF BW (Interleaved)	4000 MHz
Size, Weight	254 x 167 x 35 mm^3, 2.5 kg
Digital controls, Power supply, connector	TTL controls, DC +12V & -12V, Micro D 51 pin
Digital controlled Attenuation, step size	Each spilt RF band range: 31 dB, 1dB Step
Control lines, TTL	10 lines per channel, Total 40 control lines.
Integrated LO frequencies & Phase Noise	C Band, -100 dBc/Hz @ 10 kHz

Table 1 is specifications with regard to architecture and implementation methodology. Here architecture is proposed for four channels down conversion to IF (extendable to multi channels). For four channels, there are twelve RF paths with associated control lines & power supply and on the LO side, Phase Locked Loop (PLL) based LOs generation with associated programming & power supply, Eight LOs to be distributed to eight paths across four channels. Four down conversion channels, LOs generation & distribution and Digital & power supply sections as mentioned above are all implemented in compact size of 254x167x35 mm^3 with one RF input & two IF output ports, three RF/IF ports per channel. For n channels 3n ports, in quad case four inputs, eight outputs total twelve RF/ IF ports. SMA connectors are to be on single side as shown in Figure 2. Control lines for step attenuator in each split RF Band, 5 control lines per attenuator, three attenuators per channel. L Band RF signal controls are shared with C Band RF Signal, 10 control lines per channel. Total 40 attenuation controls & 1 switch control and power supply are to be routed internally to all channels. All these are input through Micro D - 51 pin connector on the other side as shown in Figure 3.

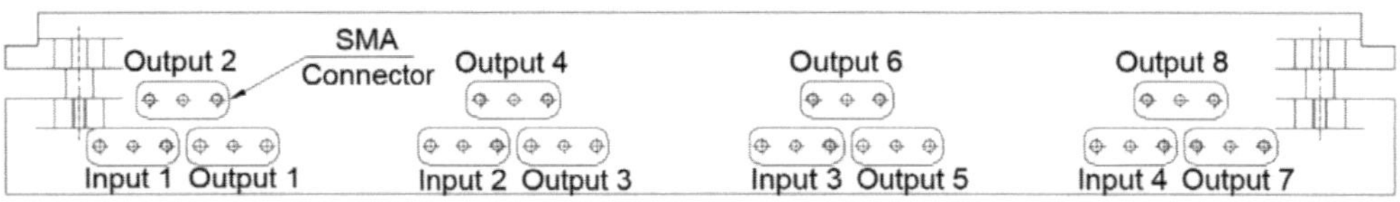

Figure 2: Module RF side view.

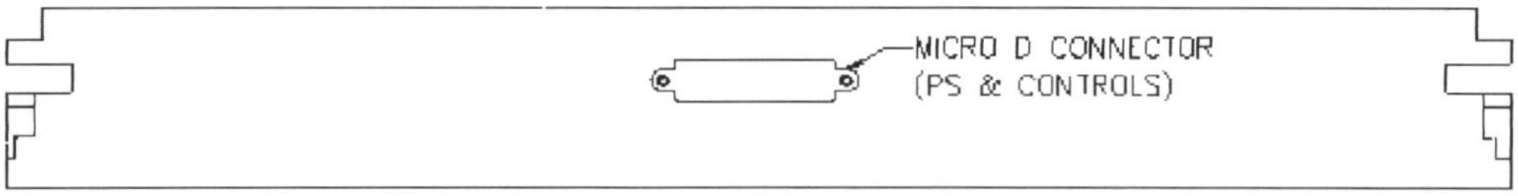

Figure 3: Module PS & control side view.

2. Module Architecture Planning

Module needs to accommodate quad channels. The module architecture design should achieve specifications as mentioned in Table 1, other electrical specifications were compiled as discussed in [2]. RF chains single plane implementation enables easy implementation; ensure reliable performance, faster reproducibility. For four channels, RF input separations into three bands, RF to IF conversions with LO implementation on a single plane is as shown in Figure 4. From the figure as observed number of RF to LO overlap observed are four, LO to LO overlaps observed are 6. In total ten overlaps were estimated.

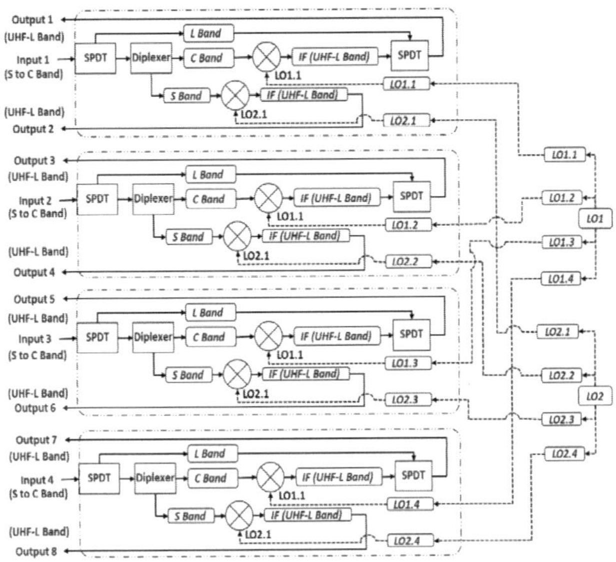

Figure 4: Frequency conversion architecture single plane implementation.

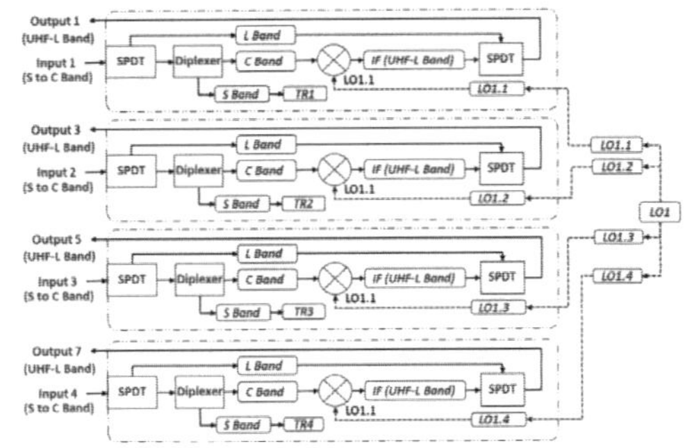

Figure 5: Frequency conversion architecture Plane 1 implementation

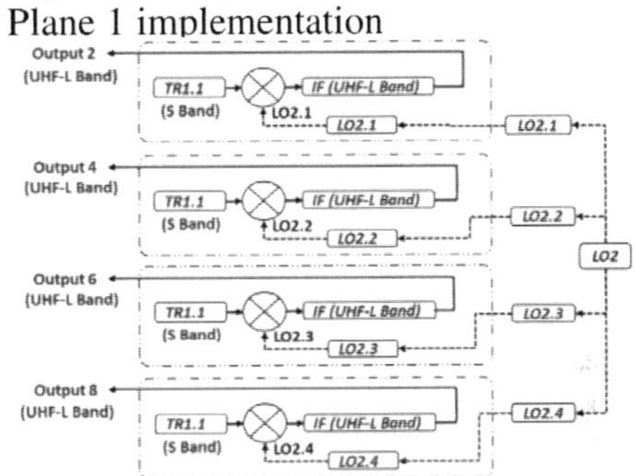

Figure 6: Frequency conversion architecture Plane 2 implementation

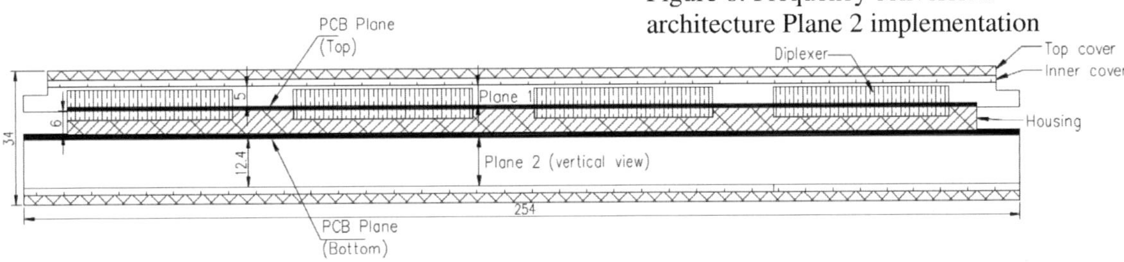

Figure 7: Vertical cross section (Connector side) of module

Two plane implementation was planned to overcome overlaps problem. Module was divided into two planes in vertical direction as seen from connector side as shown in Figure 7. Planes are parallel to each other. Plane 1 was for RF input separation into split band 1, 2 & 3 (L, S & C Bands), post separation, Spilt band 1 amplification with filtering & Spilt band 3 amplification, filtering and down conversion to IF & respective LO generation. Separated spilt band 2 was transferred to other plane. Plane 2 contained spilt band 2 reception from other plane, amplification, filtering and down conversion to IF, associated LO generation and distribution, 100 MHz TCXO reference distribution to both PLLs, microcontroller & associated circuitry, attenuator control lines distribution to dual bands for four channels. Figure 5, 6 shows Plane 1, 2 RF & LO chains implementation. Number of overlaps are reduced to zero. Multilayer composite PCBs[4] were planned due coexistence of RF/ control lines & power supply. PCBs in planes 1 & 2 are facing away from each other, each facing its respective Top cover. Spilt band 1 was transferred from Plane 1 to Plane 2 using coaxial based vertical transition [3]. Control lines & power supply from plane 2 were transferred to plane 1 using a board to board connector. In each channel Instantaneous IF Bandwidth of 4 GHz on two ports 2 GHz each present simultaneously, was achieved using Diplexer [1]. Diplexer was sized to 45x42x7.5 mm^3. Diplexer placement per channel is shown in Figure 7.

3. Local Oscillator

LO frequencies chosen to be in C Band, IF is in UHF-L Band, hence LO leakage into output & image rejection requirements are easily met. Two Local oscillators LO1 & LO2 are generated & are used to down convert spilt bands 2 & 3 to IF. LO1 & LO2 are derived using PLL & VCO architecture referenced to crystal oscillator [4]. Phase Noise requirement of -100 dBc/Hz at 10 kHz offset, are to be met. Reference oscillator Phase Noise requirement was arrived as per calculations below.

Using 100 MHz reference: at 10 kHz offset; Ref PN < -100-20log(7000MHz/100MHz);
Ref PN < -136.9 dBc/Hz
Reference 100 MHz TCXO chosen with phase noise specifications of -135 dBc/Hz at 1 kHz, -145 dBc/Hz at 10 kHz, -150 dBc/Hz at 100 kHz offset. With the TCXO selected we have 10 dB margin in Phase Noise for practical considerations.

PLL Phase Noise estimation is done based on Floor FOM, Flicker FOM, VCO Frequency output f_{vco}, Phase detector input frequency $f_{pd,}$ offset frequency at wich Phase Noise is measured $f_{offset.}$

PLL was simulated using TI's PLLatinum sim tool. Loop filter was tuned for optimised bandwidth to obtain required Phase Noise specifications. PLL simulation carried out is as shown in Figure 8. Phase Noise simulation results -96.7 dBc/Hz at 1 kHz offset, -106.6 dBc/Hz at 10 kHz offset, -106.9 dBc/Hz at 100 kHz offset was achieved. Measured Phase Noise simulation results are as shown in Figure 9, required specifications were achieved.

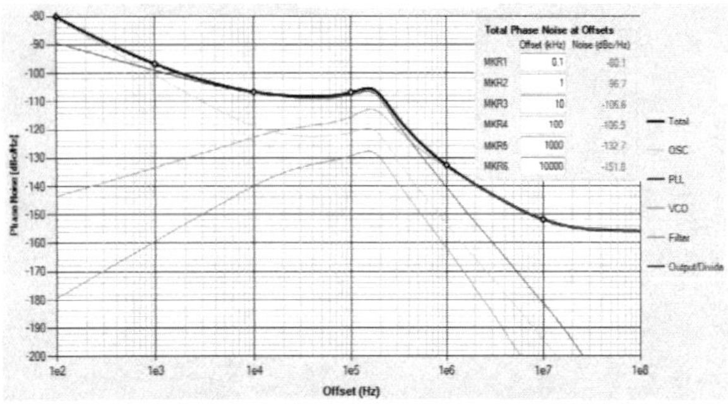

Figure 8: Phase Noise Simulation

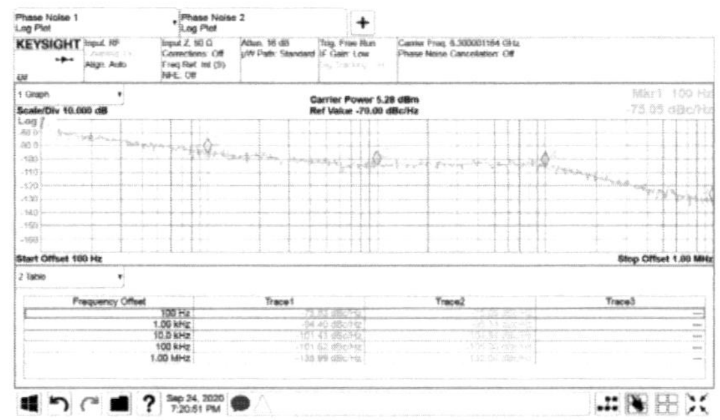

Figure 9: Measured Phase Noise.

4. Conclusion

Model architecture discussed above was developed for four channels. Architecture discussed is scalable in nature it can be scaled down or up. Upon scaling length 167 mm, height 35 mm remains constant, width of module changes as per no of channels. Test results complied to all the required specifications. Unit was integrated into main EW system, and it achieved system requirements.

5. Acknowledgement

The authors wish to acknowledge the assistance and support of AGM Super Components & all the members of Super Components PD&IC during the design phase. And also D&E EW and assembly division of Microwave Super Components, BEL Bengaluru during product realization.

6. References

[1] Hsiao-Lan Chan; Ching-Her Lee; Chung-I G. Hsu. "Balanced dual-band diplexer design using microstrip and slot-line resonators" 2015 Asia-Pacific Microwave Conference (APMC)

[2] Madiwalesh M P; Gaurav A; Rohit L "Quad channel wide band frequency conversion with integrated LO for high IF Bandwidth" 2020 IEEE international RF & Microwave conference 2020 (RFM2020)

[3] Gaurav Anand; Rohit Lahiri; Rahul Sadhu "Wide Band Microstrip to Microstrip Vertical Coaxial Transition for Radar & EW Applications" Asia-Pacific Microwave Conference 2016 (APMC)

[4] Madiwalesh M. P; Ruchit M. S; Harikrishna M. V "Design & Development of Agile C-Band Synthesizer for RADAR" 2018 IEEE MTT-S International Microwave&RF Conf (IMaRC)

Design of enhanced gain patch antenna at 2.4 GHz and 3.5 GHz for wireless application

Mohammad Addad Azeman[1], Norun Abdul Malek[1], Sarah Yasmin Mohammad[1] and Mohd Ifwat Mohd Ghazali[2]

[1] Microwave, Communication and Information System Engineering (MCISE RG), Kulliyyah of Engineering, International Islamic University Malaysia, Jalan Gombak, 53100 Kuala Lumpur, Malaysia

[2] Smart RG, University Sains Islam Malaysia, Bandar Baru Nilai, 71800 Nilai, Negeri Sembilan, Malaysia
Email: addadazeman@gmail.com, norun@iium.edu.my, smohamad@iium.edu.my, ifwat.ghazali@usim.edu.my

Abstract. In this work, the analysis of patch antenna design at the resonant frequency of 2.4 GHz and 3.5 GHz are presented. Both proposed antennas made of RT Duroid 5880 are designed and simulated using CST Studio Suite software. The antennas are designed, and their performances are examined. The antennas have been designed based on the size improvement technique to improve the fabrication tolerance. The characteristics of both antennas' including return loss, radiation pattern, surface current and directivity or gain were compared. The optimization of patch measurement to achieve the desired resonant frequency were also displayed.

1. Introduction

The implementation of microstrip patch antenna in wireless communication is popular due to its good performance despite having a low-profile configuration [1]. However, it subjected to experience a low gain and narrow bandwidth [2]. It is very crucial for a patch antenna to exhibit a low return loss (≤-10 dB) while having a high gain or directivity to attain excellent communication. Other than executing an array configuration, the gain of a patch antenna can also be amplified by increasing its dimensions [2].

Thus, in order to maximize the performance, the improvement design of microstrip patch antenna gain and bandwidth at high frequency is imitated and optimized for the ISM frequency band of 2.4 GHz. Moreover, the method is also enhanced to adapt the mid band frequency of 5G application at 3.5 GHz. The frequency band are favoured as the higher frequencies are more prone to precipitation losses due to weather [3,4]. In this research, the RT Duroid 5880 substrate is used for both design because of its excellent permittivity and conformal ability.

2. Antenna Design

These patch antennas are designed using RT Duroid 5880 substrate with a dielectric constant, $\varepsilon_r = 2.2$ [5] and thickness, t =1.57 mm at 2.4 GHz and 3.5 GHz frequency. In this research, the patch antennas will initially be designed according to size extension method, and it is optimized to obtain the desired resonant frequency and performance. Figure 1 shows the proposed CST antenna design. It consists of two patches, upper patch and lower patch connected by a microstrip line, ground plane and a microstrip feedline. This work is taken from [2] but modified at lower frequency range of 2.4 and 3.5 GHz.

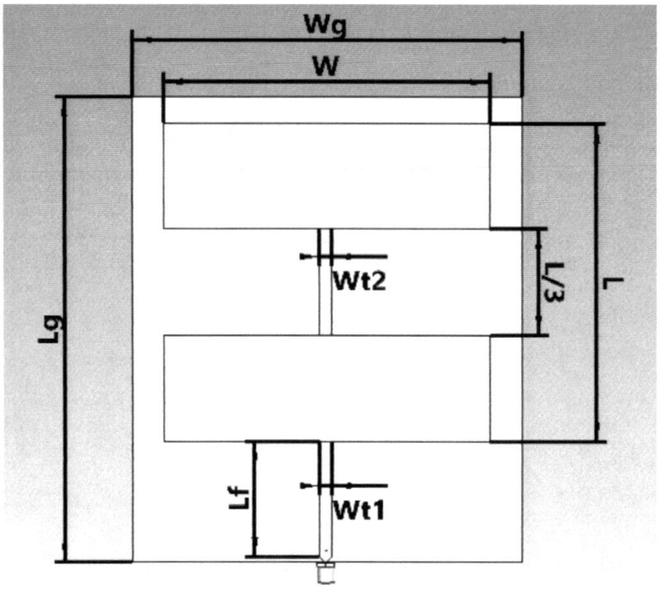

Figure 1. Patch antenna configuration (Top view)

The radiating patch width 'W' and length 'L' can be expressed as (1-2) [6]

$$W = \frac{(2N+1)}{\sqrt{(\varepsilon_r+1)/2}} \times (\frac{\lambda 0}{2}) \quad (1)$$

$$L = \frac{(2N+1)}{\sqrt{\varepsilon_{eff}}} \times \left(\frac{\lambda}{2}\right) - 2\Delta L \quad (2)$$

Where N is non-negative integer (N=1), $\lambda 0$ is the operating wavelength and ε_r is the relative dielectric constant. The effective dielectric constant 'ε_{eff}' and 'ΔL' value is obtained through following formulas (3-4) [7]

$$\Delta L = 0.412h \frac{(\varepsilon_{eff}+0.3)(0.264+\frac{W}{h})}{(\varepsilon_{eff}-0.258)(0.8+\frac{W}{h})} \quad (3)$$

$$\varepsilon_{eff} = \frac{\varepsilon_r+1}{2} + (\frac{\varepsilon_r-1}{2})(1 + 12\frac{W}{h})^{-1/2} \quad (4)$$

For this antenna structure, the width of upper and lower patch connector is set to be equal to the width of the feedline ($W_{T2} = W_{T1}$). The ground length 'Lg' and ground width 'Wg' were set to Lg=L+12h and Wg=W+12h respectively. The length of the patch 'L' is optimized to ensure the designed antennas resonate at the desired frequency, 2.4 and 3.5 GHz. The width of patch 'W' and width of feedline 'Wf' are improvised to minimize the return loss. The final optimized parameters are shown in table 1. The fabricated antennas is shown in figure 2. Figure 3 displays the experimental setup to measure the characteristics of the antennas.

Table 1. Design Parameters for antennas at 2.4 GHz and 3.5 GHz

	Dimensions (mm)	
Parameter	**2.4 GHz**	**3.5 GHz**
Length of patch (Lp)	119.5	80.3
Width of patch (Wp)	123.23	90
Length of ground plane (Lg)	175.19	134.586
Width of ground plane (Wg)	147.07	108.84
Width of substrate (h)	1.57	1.57
Thickness of radiating patch (t)	0.035	0.035
Length of feedline (Lf)	45.69	44.286
Width of feedline (Wf)	4.837	4.837

Figure 2. Fabricated antennas

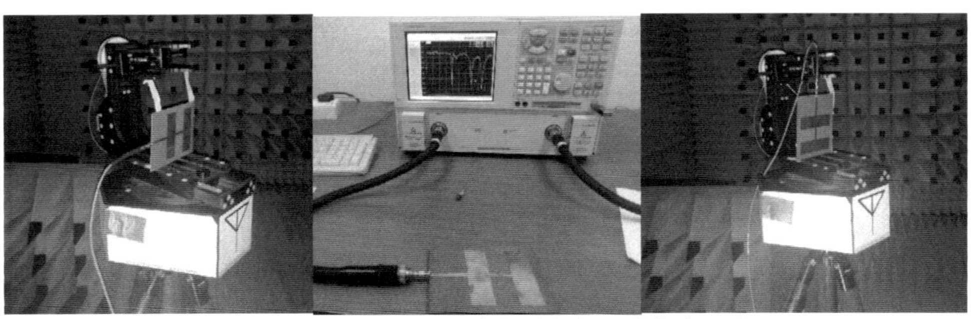

Figure 3. Experimental setup of the fabricated designed antennas

3. Results And Discussion

The return losses of 2.4 GHz and 3.5 GHz patch antenna are shown in figure 4. Based on the measured result, the resonant frequency of the 2.4 GHz antenna shifted by 20 MHz from the simulated result. The measured antenna's return loss is -13.23 dB, 8% higher compared to the simulated S11. Apart from that, the measured antenna's bandwidth is 29.4 MHz, which is 8% lower than the simulated results.

The measured 3.5 GHz patch antenna resonates at 3.55 GHz, shifted by 50 MHz from the simulated resonant frequency. The antenna's return loss is almost identical between the measured and simulated results which is around -14 dB. Meanwhile, the measured antenna's bandwidth is 50 MHz, which is 4% higher than the simulated results. The shift in resonance frequency might happen due to fabrication tolerance during the manufacturing process [8]. Both antennas exhibit low return losses across resonant frequency, ≤ 10 dB which is relevant for patch antenna [9].

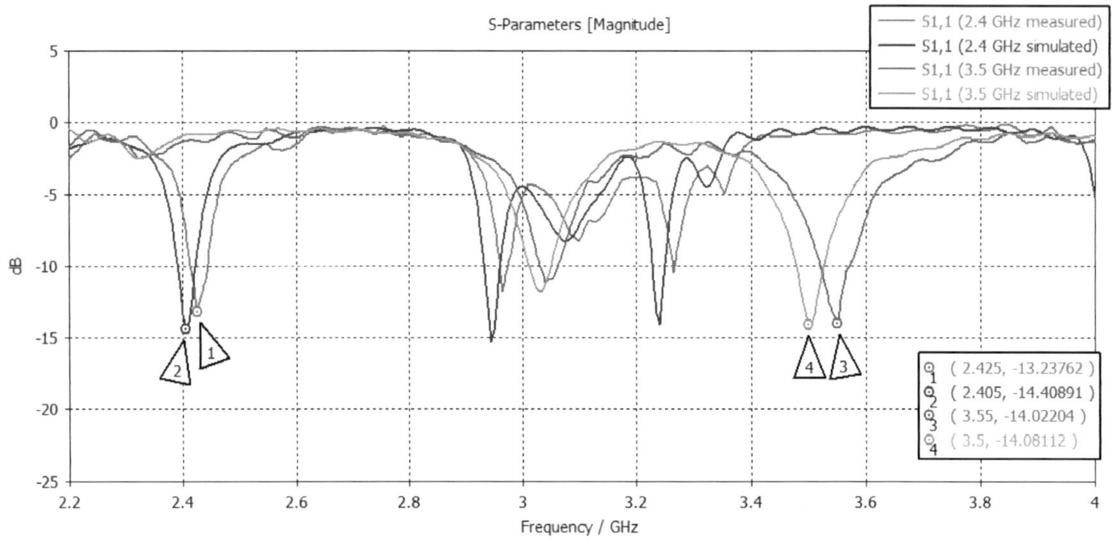

Figure 4. Simulated and measured S11 of 2.4 GHz and 3.5 GHz patch antenna

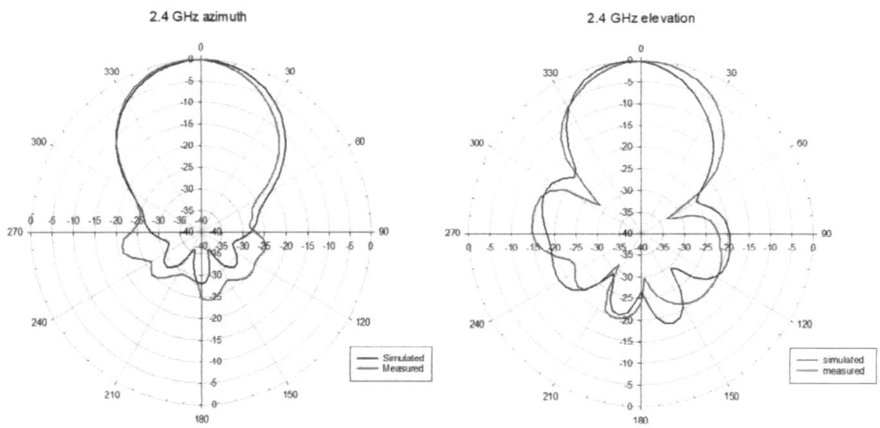

1

(a) (b)

Figure 5. Simulated and measured (a) H-plane and (b) E-plane pattern at 2.4 GHz

Figure 5 shows simulation and measured results of far-field patterns in both E and H planes of 2.4 GHz antenna. From simulation, the gain simulated is 12.41 dB. Meanwhile, the measured gain for the antenna is close to 12.2 dB or 1.67% lower compared to simulated gain.

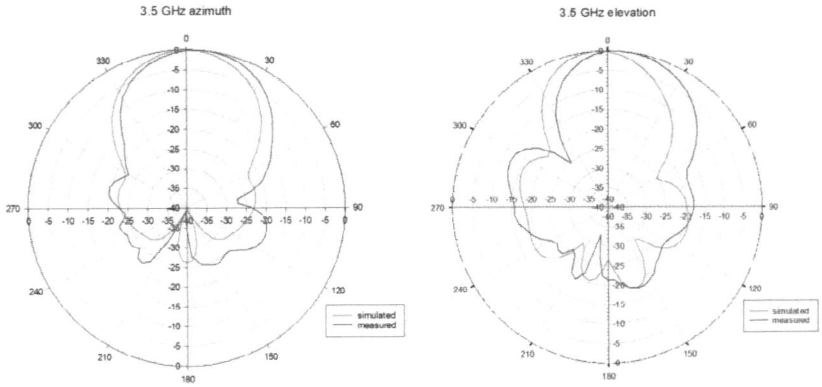

(a) (b)

Figure 6. Simulated and measured (a) H-plane and (b) E-plane pattern at 3.5 GHz

Figure 6 shows simulation and measured results of far-field patterns in both E and H planes of 3.5 GHz antenna. The gain simulated for 3.5 GHz patch antenna is 13.14 dB. As for the measured result, the gain obtained is around 13.03 dB, less than 1% difference to the simulated gain.

The gain comparison method is applied to measure the patch antenna gain by using broadband horn antenna as the reference. This antenna can operate from 1 GHz to 18 GHz. The gain of reference horn is shown in figure 7. The gain of patch antenna is calculated using the formula (5) [7]

$$G_M = G_R + \Delta G \tag{5}$$

where G_M is the gain of measured antenna, G_R is the gain of reference antenna and ΔG is the difference between them. Based on the figure 7, the standard gain at 2.4 GHz and 3.5 GHz horn antennas are around 13.7 dB and 13.5 dB respectively.

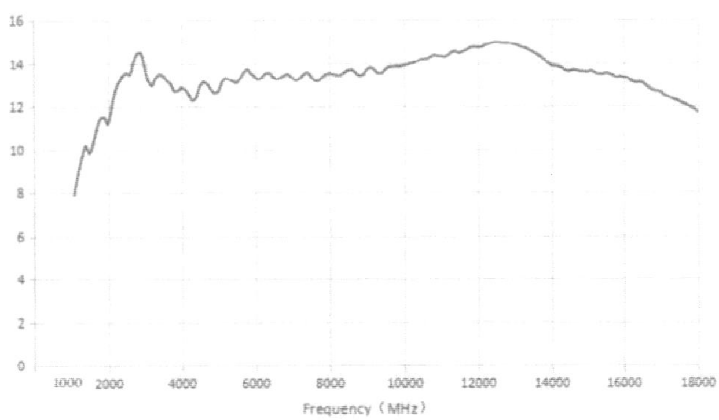

Figure 7. The gain of standard horn antenna [10]

The measurement of antenna gain is presents in table 2 with the comparison to the simulated antenna gain. Overall, the performances of the design antennas are concluded in table 3.

Table 2. The comparison table between simulated and measured gain

Antenna types	Max patch power density measured	Max horn power density measured	Difference between horn and patch	Gain measured (dB)	Gain simulated (dB)
2.4 GHz	-33.299	-34.8	1.501	12.2	12.41
3.5 GHz	-34.848	-35.315	0.467	13.03	13.14

Table 3. Design antennas' performances

Antenna types	Bandwidth (MHz)	Gain (dB)	S11 (dB)	Radiation Efficiency (%)
2.4 GHz	29.4	12.2	-13.21	96.7
3.5 GHz	50.0	13.03	-14.02	97.9

4. Conclusion
Both patch antennas were designed and simulated using CST Studio software. The simulated results of the two different antennas resonate at 2.4 and 3.5 GHz were analyzed and compared to find out the performances in terms of return loss, radiation pattern and surface current. It was observed that the characteristics obtained at 3.5 GHz frequency does not have significant difference to 2.4 GHz antenna despite having a smaller overall patch size. The antennas will be fabricated to verify the simulated results. In addition, the antenna design will be further fabricated using 3D printing using PLA or ABS substrate to be applied to other wireless communication purposes.

References
[1] S. Palanivel Rajan and C. Vivek, "Analysis and design of microstrip patch antenna for radar communication," *Journal of Electrical Engineering and Technology*, vol. 14, no. 2, pp. 923–929, Mar. 2019, doi: 10.1007/s42835-018-00072-y.
[2] M. S. Rabbani and H. Ghafouri-Shiraz, "Improvement of microstrip patch antenna gain and bandwidth at 60 GHz and X bands for wireless applications," *IET Microwaves, Antennas and Propagation*, vol. 10, no. 11, pp. 1167–1173, Aug. 2016, doi: 10.1049/iet-map.2015.0672.
[3] J. Monica and P. Jothilakshmi, "A Design of Bandwidth-Enhanced Conformal Antenna for Aircraft Applications," *IETE Journal of Research*, 2020, doi: 10.1080/03772063.2020.1829507.
[4] J. Monica and P. Jothilakshmi, "Gain enhanced conformal patch antenna with defected ground for aircraft applications," *IET Communications*, vol. 14, no. 2, pp. 290–299, Jan. 2020, doi: 10.1049/iet-com.2018.6042.
[5] "RT/duroid ® 5870 /5880 High Frequency Laminates Some Typical Applications: • Commercial Airline Broadband Antennas • Microstrip and Stripline Circuits • Millimeter Wave Applications • Military Radar Systems • Point to Point Digital Radio Antennas," 2017.
[6] S. Baudha BITS Pilani, K. K. Birla Goa, R. Singh Rathore, S. Baudha, and S. Pandey, "A 2.4 GHz Microstrip Patch Antenna with a Single Slot for WLAN Application," 2012.
[7] C. A. Balanis, *Modern antenna handbook*. Wiley, 2008.
[8] M. Ramadan and R. Dahle, "A flexible 2.4 GHz microstrip patch antenna using a 3-D printed tile array design," *2017 IEEE MTT-S International Microwave Workshop Series on Advanced Materials and Processes for RF and THz Applications, IMWS-AMP 2017*, vol. 2018-Janua, no. September, pp. 1–3, 2018, doi: 10.1109/IMWS-AMP.2017.8247424.
[9] K. R. Kashwan, V. Rajeshkumar, T. Gunasekaran, and K. R. Shankar Kumar, "Design and characterization of Pin fed microstrip patch antennae," in *Proceedings - 2011 8th International Conference on Fuzzy Systems and Knowledge Discovery, FSKD 2011*, 2011, vol. 4, pp. 2258–2262. doi: 10.1109/FSKD.2011.6020028.
[10] "Broadband Horn Antenna". EC Mirowave, 2017.

Acknowledgements
The authors would like to acknowledge the Ministry of Education (MoE) Malaysia and Department of Electrical and Computer Engineering, Kulliyyah of Engineering, International Islamic University Malaysia (IIUM) for supporting this work of antenna design and testing. This research is supported under grant FRGS/1/2018/ICT03/UIAM/02/2. We would also like to thank Malaysia-Japan International Institute of Technology (MJIIT) for providing facilities to measure the radiation pattern and gain of the designed antenna.

Cylindrical Dielectric Lens Antenna for 5G Mobile Base Station

F Ansarudin[1,2*], Y Yamada[3*] and K Kamardin[3]

[1]Faculty of Engineering & Built Environment, Universiti Kebangsaan Malaysia, Selangor 43650, Malaysia.
[2]School of Electrical Engineering, Universiti Teknologi Malaysia, Johor 81310, Malaysia.
[3]Department of Electronic Systems Engineering, Malaysia-Japan International Institute of Technology, Universiti Teknologi Malaysia, Kuala Lumpur 54100, Malaysia.

*Corresponding authors' email: yoshihide@utm.my; farizah.ansarudin@ukm.edu.my

Abstract. A wide range new technologies such as millimetre wave, massive MIMO technology, multi beam base station antenna are required in the fifth-generation (5G) mobile system. For multi beam base station application, a dielectric lens antenna is selected due to capabilities of achieving excellent radiation patterns in a wide angular range by applying the lens shaping method. A cylindrical structure is suitable for a lens antenna installation on the base station pole. In this paper, a cylindrical dielectric lens antenna is proposed and multi beam radiation patterns are investigated through the calculations by an electromagnetic simulator (FEKO) at the operating frequency of 28 GHz. As a result, good multi beam radiation patterns are achieved in the scanning angle from 0°to 30° with total gain reduction around 5dBi and antenna efficiency of about 95.9%.

1. Introduction

Nowadays, communication system which providing fast rate transmission, low latency, extremely high traffic volume density, super-dense connections and cost efficiencies are the main requirement for the deployment of 5G mobile system. At the 5G mobile, new features such as millimetre wave operation, small cell size and multi beam base station antenna to meet massive multiple-input multiple-output (MIMO) requirements [1]. To ensure high quality of services, high performance of multibeam scanning for 5G base station is required. The current microwave frequency spectrum suffers from congestion and are not able to support the demand for higher data rate mobile smartphone users due to the limitation of spectrum at these frequencies. In developing new base station antennas, multiple radiating beams with high gain performance are expected for the 5G base station as one radiating beam is assigned for one user as shown in figure 1(a). Therefore, the base station antenna size is expected to be less than 30 cm at the mm-wave operation.

According to [2-4], the dielectric lens antenna is known to produce excellent multi beam with very low network losses, simple structure and easy to fabricate. Based on previous studies reported by [5-6], numerous methods for non-uniform of Luneburg lenses have been applied for the use of communication applications and its feeding networks. Nevertheless, the capabilities of performing wide angular scanning for multi beam applications were not discussed in detail in those works. Ray tracing method based on the Geometrical Optics (GO) [7-8] was employed for shaping lens surfaces in achieving excellent radiation patters, where lens shaping is conventionally analysed by the aperture distribution method or Abbe's sine method [2]. For the base station pole installation, cylindrical dielectric lens antenna (CDLA) structure is suitable as shown in figure 1(b). In this paper, a cylindrical dielectric lens antenna is designed and the performance of multi beam characteristics for scanning beam from 0° to 30° at the vertical plane is observed.

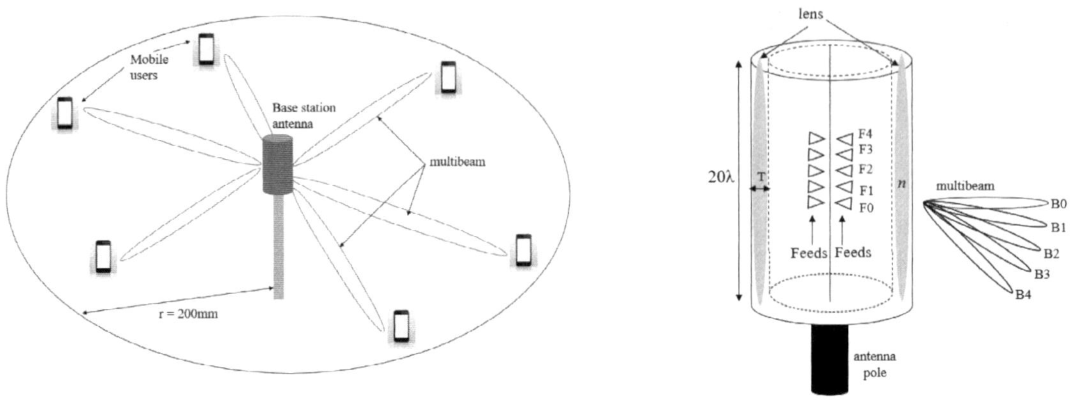

(a) (b)

Figure 1. (a) Concept of multi beam 5G base station antenna (b) A cylindrical lens structure for the pole installation

2. Antenna Configuration and Lens Shaping Method

The principle of lens shaping is based on Snell's Law for constructing a surface 1 (S1) and surface 2 (S2), respectively. The lens design equations and function of equations are expressed all in table 1 to shape the lens structure by ray tracing method. The new design method is proposed to shape a small and this lens curvature, called as Straight-Line Condition (SLC) [9]. The simultaneous differential equations composed of Eq. (1), (2) and (5) are solved by a MATLAB program. Figure 2 shows the lens antenna configurations comprises of two surfaces, S1 and S2, respectively. The lens is symmetrically rotate at the z-axis the x-axis corresponds to the radial direction of the lens. The radiation pattern of the horn is indicated as $E_p(\theta)$ and the illumination distribution on the antenna aperture plane is shown by $E_d(x)$. The ray emitted from feed at maximum angle, θ_M reaches S1 is indicated by (r,θ). Then, the refracted ray at S1 produced by the angle, ϕ, reaches S2 at the point indicated by (z,x) and paralled to aperture plane. The shaped SLC lens by MATLAB program is shown in figure 3. From the figure, the rays are radiated from the feed point and formed to be parallel rays to the z-axis by the effect of lens refractions.

Table 1. Lens design equation

Subject	Equation		Function of equations
Snell's Law on S1	$\frac{dr}{d\theta}=\frac{nr\sin(\theta-\phi)}{n\cos(\theta-\phi)-1}$	(1)	r is determined for θ change
Snell's Law on S2	$\frac{dz}{d\theta}=\frac{n\sin\phi}{1-n\cos\phi}\frac{dx}{d\theta}$	(2)	z is determined for θ change
Constant electrical path length condition	$L_t=r+\frac{n(Z-r\cos\theta)}{\cos\phi}+Z_o-Z$	(3)	ϕ is given by θ
Straight-Line Condition (SLC)	$x=L\tan\theta$	(4)	$dx/d\theta$ in eq. (2) is given and
	$\frac{dx}{d\theta}=L\sec^2\theta$	(5)	x is determined for θ change

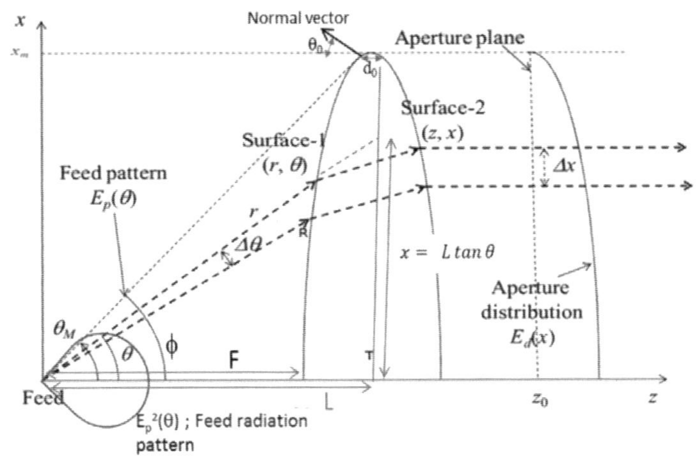

Figure 2. Lens antenna structure

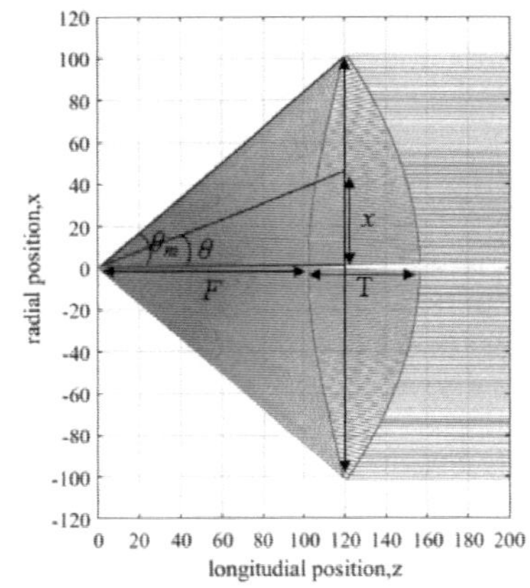

Figure 3. Designed SLC lens shape

3. Electromagnetic Simulation Model

In order to provide antenna for a base station, practical installation needs to be taken into account such as economically, simple structure, high performance and reliable. In this paper, cylindrical structure is designed at the ratio of focal length to diameter (*F/D*). The electromagnetic simulator FEKO is employed. Simulation structural parameters are shown in table 2. The chosen of most suitable lens material is based on practicality of easy fabrication with small ratio of *F/D* and small lens thickness. *F/D=0.6* is selected. The lens diameter is set to *D*=200mm, approximately 20λ. As for the dielectric material, the use of TPS-PC (Polycarbonate) by Toray is considered and *n*=1.7 (ε_r=2.9) is selected. The 90-degree cylindrical SLC lens geometry is shown in figure 4.

Table 2. Structural parameters

Antenna parameter	Dimension
Operation frequency	28 GHz
Diameter, *D*	200 mm
Focal Length, *F*	119.18 mm
Thickness, *T*	51.3 m
Radius of Cylinder, L	144.88 mm
Refractive index, *n*	1.7
Angle to lens edge, θ_m	40°

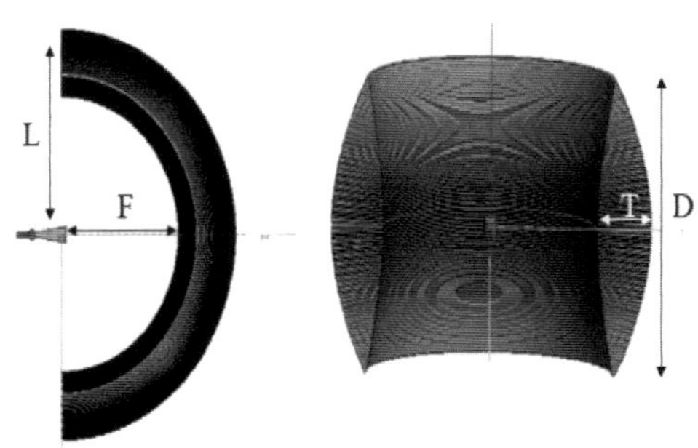

Figure 4. Cylindrical SLC antenna

4. Electric Field Distributions in the Near Field

A practical pyramidal feed horn antenna is designed with the ratio of focal length to diameter, *F/D*=0.6 to illuminate the lens edge. Figure 5 shows the horn dimension with edge level for both XZ-plane and YZ-plane of about -10.94 dB and -8.38 dB, respectively. The electric field distributions at the near field position of the lens aperture are shown in figure 6. At the position ρ=173 mm, the amplitude and the phase distribution are shown in figure 6(a). From the figure, small ripples are observed and aperture distribution of field intensity is well agreed between simulation (FEKO) and calculation in MATLAB at the lens edge, -12.9 dB and -12.21 dB, respectively. From figure 6(b), existence of standing waves is observed inside the lens. The reason of small ripples at the amplitude and phase distributions is considered to be the standing waves. For elimination of the standing waves, attachment of a matching layer is effective.

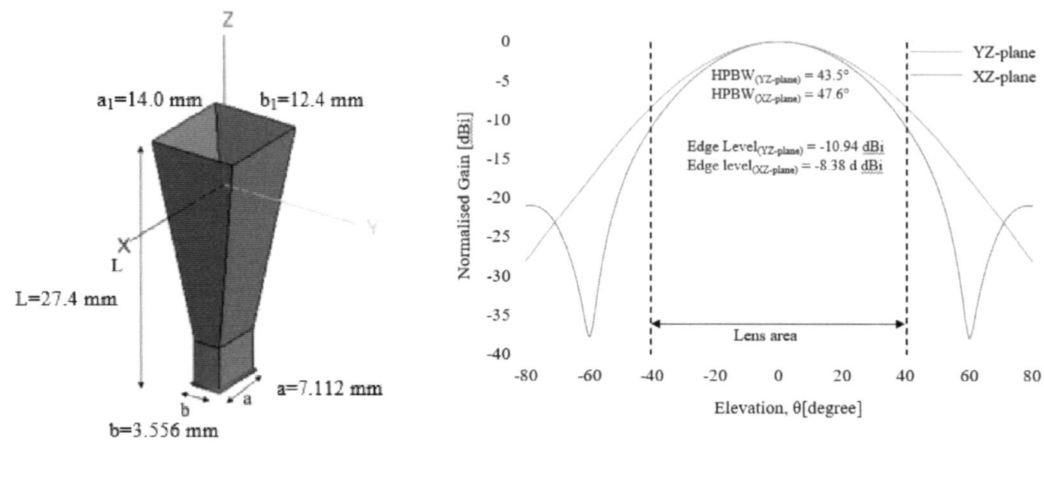

(a) (b)

Figure 5. The pyramidal horn antenna for *F/D* =0.6 (a) horn dimension (b) horn radiation pattern

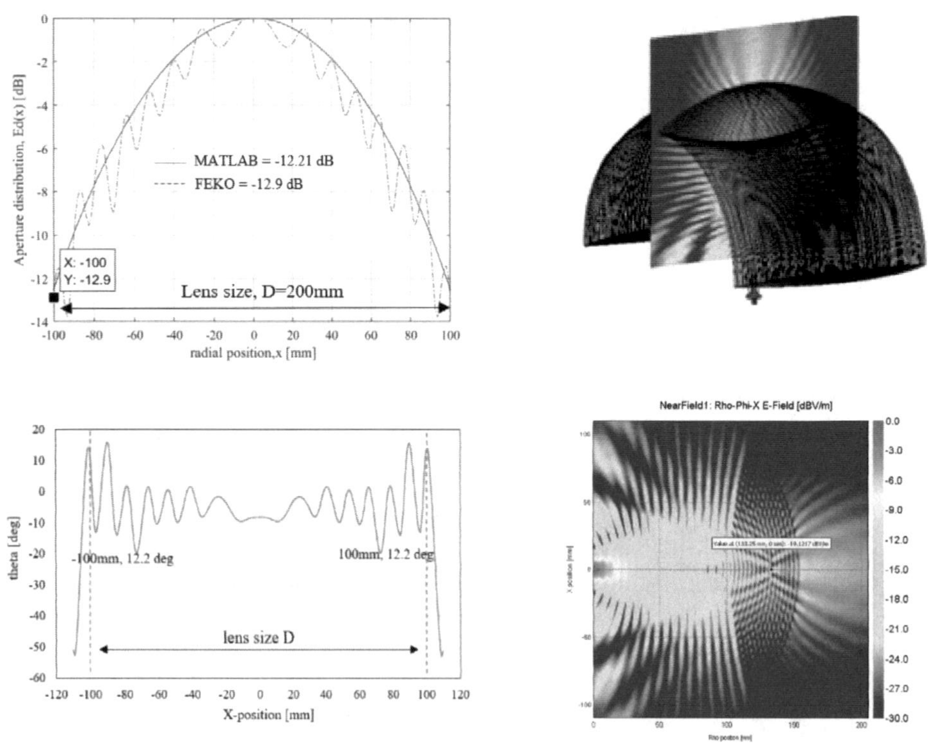

(a) (b)

Figure 6. (a) Amplitude and phase distributions (b) Electric field distribution

5. Multi Beam Radiation Patterns

As a fundamental data, the radiation characteristics at the on-focus feed condition is examined as shown in table 3. Gain values of MATLAB and FEKO agree well with the theoretical value with antenna efficiency around 95.9%. The calculated beam width by analytical and simulation are similar and slightly larger than the theoretical value.

Table 3. Calculated and simulated HPBW and gain for on-focus beam

Method	HPBW [θ_{3dB}]	Gain [dBi]
Theoretical	3.21°	24.86
Analytical (MATLAB)	3.21°	24.34
Simulation (FEKO)	3.28°	24.68

For multi beam performance, the feed positions are determined by the focal region ray tracing technique employed by the Ray-Launching Geometric Optic (RL-GO) solver in FEKO [10]. Figure 7 shows the the radiation patterns for on-focus feed (θ_F = 0°) and scanning angle up to θ_F = 30°. The performance of multi beam radiation characteristic is solved by the transmitting mode employed by Method of Moment (MoM) solver. It is observed at the feed angle from θ_F = 0° to 30°. The centre beam (θ_F = 0°) shows high gain value 24.68 dBi and narrow half power beam width (HPBW) θ_{3dB}=3.28° are obtained. It is clearly seen that good multi beam radiation patterns are achieved with beam direction up to θ_S=33.1°. After the beam direction of larger than 30°, the beam deformation becomes a bit large and high side lobes appear. Table 4 shows the summary of multi beam characteristic. As a result, the cylindrical SLC lens antenna is suitable for the use of 5G mobile multibeam base station due to wider scanning beam coverage at vertical plane and low reduction in gain around 3.16 dBi at θ_F=30°

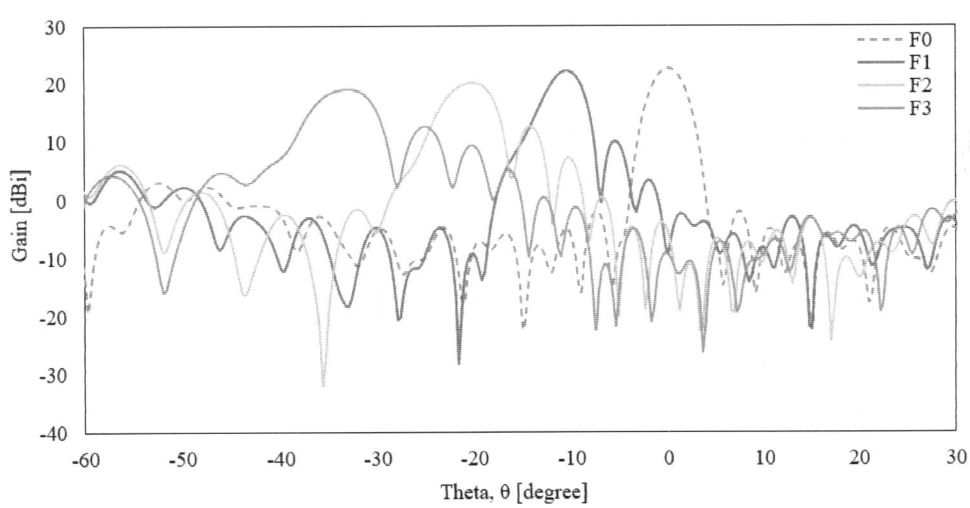

Figure 7. Multi beam radiation pattern based on locus focal region ray tracing technique

Table 4. Multi beam characteristics of cylindrical SLC lens

Feed Position	Feed Angle θ_F (°)	Beam Direction θ_S (°)	HPBW θ_{3dB}	Gain (dBi)	Gain Difference (dBi)
F0	0	0	3.28	22.68	0
F1	10	10.5	3.49	22.11	-0.57
F2	20	20.1	3.96	20.24	-1.43
F3	30	33.1	5.23	19.20	-3.16

6. Conclusion

The cylindrical lens antenna with a shaped lens surface is proposed. The lens shape is designed based on the straight-line condition (SLC). The vertical cylindrical structure with the SLC shaped cross section is simulated by the MoM solver for electromagnetic simulations. By shifting feed positions from θ_F=0° to θ_F=30°, multi beam radiation patterns are calculated. As a result, the total gain reduction around 5 dBi and good multi beam shapes are achieved for wide scanning angle. The useful of a new stucture of cylindrical SLC lens antenna is suitable for 5G multi beam base station with coverage beam up to θ_S=33.1°.

References

[1] Rappaport, T.S et al. *Millimeter wave mobile communications for 5G cellular: It Will Work!,* IEEE Access, 2013. 1: p. 335-349.
[2] Lo, Y.T and Lee, S.W. *Antenna Handbook*. Van Nostrand Reinhold, Ch. 16. Lens Antenna, 2013. pp.19-23.
[3] Thornton, J and Kao-Cheng. *Modern Lens Antenna for Communications Engineering*. 2013. John Wiley & Sons.
[4] Tajima, Y. and Yamada,Y. *Design of Shaped Dielectric Lens Antenna for Wide Angle Beam Steering*. Electronics and Communications in Japan. Part 3. 2006. 89(2), p. 286-296.
[5] Li, Y. et al., *Multibeam 3-D-Printed Luneburg Lens Fed by Magnetoelectric Dipole Antennas for Millimeter-Wave MIMO Applications*. IEEE Transsaction and Antenna Propagation. 2019, 67, 2923–2933.
[6] Liang, M., et al., *A 3-D Luneburg Lens Antenna Fabricated by Polymer jetting rapid prototyping*. IEEE Trans. Antennas Propagation, 2014, 62, 1799- 1807.
[7] Zhou, X.Y et al., *Three-dimensional large-aperture lens antennas with gradient refractive index*. Science China Information Science, 2013, 56, p. 1–12.
[8] Yunxiang, Z. et al., *Numerical Analysis of Dielectric Lens Antennas Using a Ray-Tracing Method and HFSS Software*. IEEE Antennas Propag. Mag. 2008, 50, 94–101.
[9] Farizah A. et al., *Multi Beam Dielectric Lens Antenna for 5G Base Station*. Sensors, 2020. 20(20), 5849.
[10] Farizah A. et al. *Focal Region Ray Tracing of Dielectric Lens Antenna for Multi Beam Applications*. 7th International Conference on Space Science and Communication, Kuala Lumpur, Malaysia (virtual conference), 23-24 Nov 2021 (accepted).

Acknowledgments

This work is carried out with the financial support from Ministry of Higher Education (MoHE) through the Fundamental Research Grant Scheme (FRGS/1/2018/TK04/UTM/01/4) and the Universiti Teknologi Malaysia Prototype Research Grant (UTMPR/PY/2019/02734).

Near Field on Inductor Loops for Inductive Couple Fed Antenna

N Q Azmi, N A Murad, N A Samsuri, A R Tanvir and A A Z A Nizam
Advanced RF and Microwave Research Group (ARFMRG), School of Electrical Engineering, Faculty of Engineering, Universiti Teknologi Malaysia (UTM), 81310, UTMJB, Johor, Malaysia.

Email: Nqamarina5@utm.my

Abstract. This paper presents an evaluation on the near field of square loop and circular loop for inductive couple fed RFID Tag antenna. The loop is simulated with different diameter and width The turns number (N) of the loop is varies up to three number of turns for each loop length. The magnetic fields and inductance are simulated to estimate the signal strength that can be coupled to the tag antenna. To evaluate the performance, the inductive feeding loop is designed to operate at RFID UHF band (860 MHz – 960 MHz) and simulated using CST software. It is confirmed that 6 mm loop length of the square and circular loop has higher magnetic field values as the surface area of the loop is bigger. The loop with different overall dimension gives out different reactance, inductance and magnetic field, dipole antenna, CST, and loop antenna.

1. Introduction

Radio Frequency Identification (RFID) is rapidly developing technology which uses RF signals for automatic identification of objects. Now RFID finds many applications in various areas such as for inventory, electronic toll collection, supply-chain management and many more which requires long read distance. The types of RFID commonly identified based on the operational radio waves spectrum. There is low frequency (LF), high frequency (HF) and ultra-high frequency (UHF). LF (125-134 kHz) and HF (13.56 MHz) normally used for short range applications while UHF (433 MHz, 860-960 MHz) for longer range [1].

Globally, each country has its own frequency allocation for RFID. For Example, RFID UHF bands are 866-869 MHz in Europe, 902-928 MHz in North and South America, and as for in Malaysia the range is between 919 – 923 MHz as
Standard Radio System Plan (SRSP) states [2]. RFID systems comprise of interrogators and tags. The tags can be active or passive. Active tags contain an internal power source (battery) while passive tags are powered by the radio frequency energy transmitted by the interrogators.

Proper impedance match between the antenna and the chip is of paramount importance in RFID. Dipole antenna is commonly used in UHF RFID tags due to its simplicity and omnidirectional properties. The length of the total wire, which is
being used as a dipole, equals half of the wavelength (λ/2). Such an antenna is called as half-wave dipole antenna. This is the most widely used antenna because of its advantage which will give reasonable antenna length that matches the size and directivity. For the RFID chip, it will be connected to the dipole arms and a good conjugate matching between the antenna impedance, Za and the chip impedance, Zc is a must to have efficient energy transfer. There are numbers of matching techniques reported such as T-shape and inductive coupling [3].

The available loop dimension that is widely used in UHF RFID tag antenna is 6 mm x 6 mm. Therefore, this paper reports the evaluation on the square and circular loop with lower than the available loop existed which is 4 mm x 4 mm and 5 mm x 5 mm. However, when the loop length is smaller, the energy coupled to turn on the chip will be lower. Hence, the main purpose of this study is to evaluate the magnetic field and the inductance of the smaller loop and increase the flux by increasing the number of turns (N). As when the flux increases, the energy coupled will increase and also increase the read range distance.

2. Inductor loop

An inductor is basically a passive electrical component which normally built in the form of coil wire. Inductor takes the advantage of the relationship between magnetism and electricity where induced current can be produced with certain condition on magnetic field. The loop inductance calculation can be found in [4] with different shapes. In this paper, the study of square and circular loop with smaller size is chosen due to its less surface area to be occupied in the RFID tag antenna. However, from the Equation 1, it is proven that the surface area is perpendicular with the magnetic field. The smaller loop will be disadvantageous as when the magnetic field is decrease, the energy coupled to turn on the chip also will be decrease.

$$\text{Magnetic flux} = \Phi = B \times A \qquad (1)$$

As for the solution due to the lower magnetic flux, the number of loops turns will be increased to increase the flux hence increase the energy coupled. So, this paper will evaluate the square and circular shape with two and three number of turns with the proposed overall loop dimensions as mentioned.

2.1. Square loop

Figure 1 shows the square inductor loop and the respective coefficient for inductance calculation. Equation 2 shows the equation of inductance calculation.

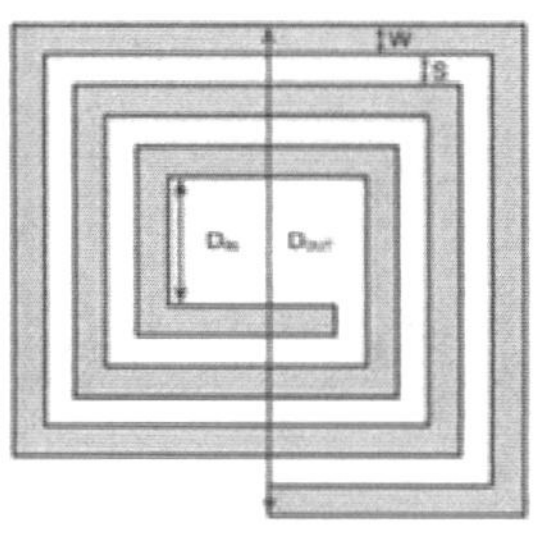

C_1	1.27
C_2	2.07
C_3	0.18
C_4	0.13

Figure 1. Square inductor loop and the coefficients [4].

Figure 1 shows the Din which is the inner diameter, Dout is the outer diameter, w is the loop width and s is the loop spacing. Dout is the parameter which will be varied from 4 mm, 5 mm and to 6 mm respectively. An inductance, L of a square inductor loop (Figure 1) can be calculated by Equation 2 below. Where N is the number of turns, μ_0 is the vacuum permeability, $4\pi\times10^{-7}$; ρ is the fill ratio, (DOUT−DIN)/(DOUT+DIN); DAVG is the average diameter, (DIN+DOUT)/2; and C_1=1.27, C_2=2.07, C_3=0.18, and C_4=0.13. Based on the equation, theoretically the inductance value is physically depending on the fill ratio and average diameter of the loop.

$$L = \frac{\mu_{0N^{2D}avgC_1}}{2}\left[ln\frac{(C_2)}{\rho} + c_{3\rho} + c_4\rho^2\right] \qquad (2)$$

2.2. Circular loop

For the circular loop, the inductance calculation is similar with the square loop. However, the coefficients of the circular loop are different as shown in Figure 2.

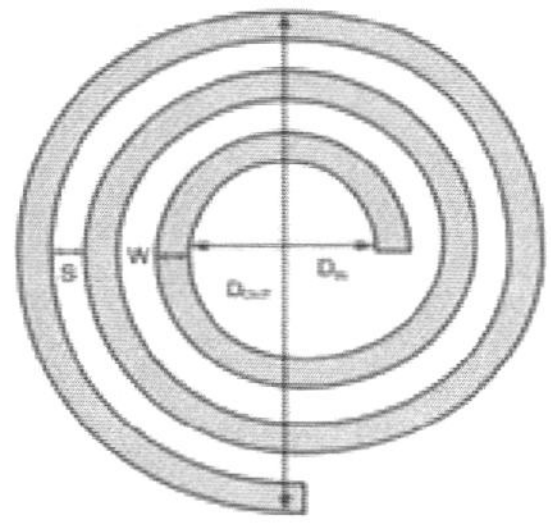

C_1	1.00
C_2	2.46
C_3	0
C_4	0.20

Figure 2. Circular inductor loop and the coefficients [4].

2.3. Inductance

The operating frequency for the RFID tag is at 920 Mhz. The inductance of the RFID tag at operating frequency can be calculated by Equation 3.

$$L = \frac{1}{(2\pi f)^2 C} = \frac{1}{(2\pi 920)^2(1.44 \times 10^{-12})} = 20.783 \times 10^{-9}\, H \tag{3}$$

Hence, the inductance of the loop must be approximately to the above inductance by emphasizing the coefficients value for both shapes in the calculation (Equation 2). The inductance calculation is calculated by PTC Mathcad. The calculation of the 4, 5 mm and 6 mm square and circle inductor loops is approximately equal to the theoretical calculation from Equation 3. On the calculation, the value of w and s is adjusted and the respective value of w and s is used in the simulation to verify the magnetic field value as shown in Figure 3 below.

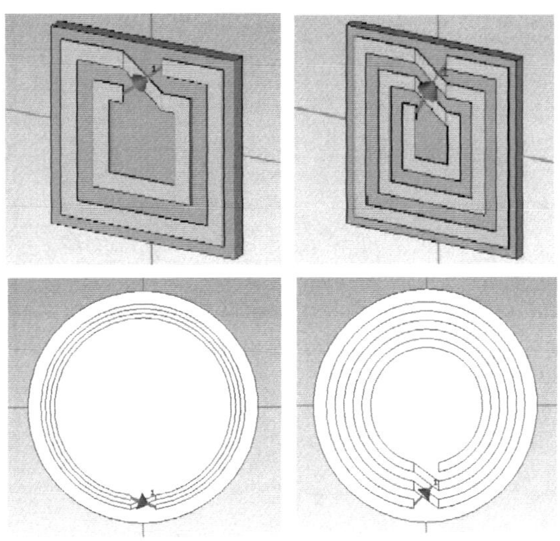

Figure 3. Simulation on CST with respective w and s values.

Table 1 and 2 will show the overall inductance for square and circular loops with LCalculated = LCalc.

Table 1. Inductance calculations of square loop

Dout / mm	Coefficient	Number of turn (N)	Loop width (w) / mm	Loop spacing (s) / mm	X_L / Ω
		SQUARE INDUCTOR LOOP			
4			0.22	0.22	118
5	C_1= 1.27	2	0.35	0.35	117
6	C_2= 2.07		0.50	0.50	113
4	C_3= 0.18		0.28	0.28	120
5	C_4= 0.13	3	0.395	0.395	120
6			0.52	0.52	119

From the previous introduction, we all know that an RFID chip can be connected to the two arms of the dipole which will be turned on for data transfer once enough energy is captured by the dipole and transferred to the feed point. A good conjugate matching between the antenna impedance, Za and the chip impedance, Zc is a must to have efficient energy transfer. The inductor loop is studied in term of the inductance value at different dimensions. Once the impedance of the loop is known, it is connected to the RFID chip. The antenna then is designed to have conjugate impedance of the loop and the chip for matching purposes at the required frequency. The Impinj Monza R6 chip will be used for the tag and there several specifications will be taken out such as the Cp and Cmount values from the Monza R6 datasheet [5]. Cp value is 1.23 pF and Cmount is 0.21 pF. Equation 4 and Equation 5 will be used to calculate the value of X_L and Xc of the loop and chip itself.

$$XL = 2\pi f \times LLoop \tag{4}$$

$$XC = \frac{1}{2\pi f \times C} = \frac{1}{2\pi(920M)(1.44p)} = j120\Omega \tag{5}$$

$$;C = Cp + Cmount$$

The calculation is continued with the inductance calculation of the circular loop and the results are shown on Table 2.

Table 2. Inductance calculations of circular loop.

Dout / mm	Coefficient	Number of turn (N)	Loop width (w) / mm	Loop spacing (s) / mm	X_L / Ω
		SQUARE INDUCTOR LOOP			
4			0.18	0.18	115
5	C_1= 1.00	2	0.3	0.3	114
6	C_2= 2.46		0.42	0.42	117
4	C_3= 0		0.25	0.25	120
5	C_4= 0.20	3	0.36	0.36	120
6			0.48	0.48	117

2.4. Reactance

The impedance of the loops also one of the vital parameters to be taken into evaluation. Each loop with different length is simulated and the reactance shows that the higher the number of loops turns, the higher the reactance values. Figure 4 evaluate that reactance is perpendicular to the number of turns for both loop shapes.

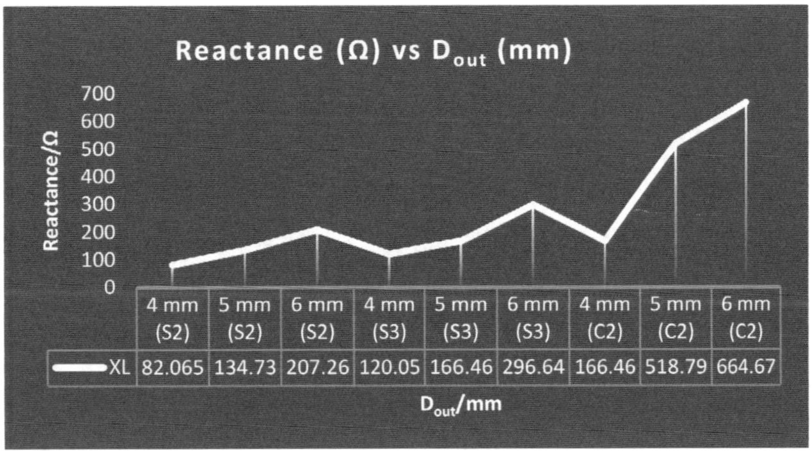

Figure 4. Reactance values of square and circular loop.

Table 3. Simulated reactance of inductor loop.

Size (mm)	N	Reactance (Ω)
4		82.065
5	2	134.73
6		207,26
4		120.05
5	3	166.46
6		296.64
4		166.46
5	2	518.79
6		664.67
4,5,6	3	Too high

2.5. Magnetic field

A magnetic field is a vector field that describes the magnetic influence on moving electric charges, electric currents and magnetic materials [6]. In electromagnetics, the term magnetic field is used for two distinct but closely related vector fields denoted by the symbols B and H. In the International System of Units, H, magnetic field strength, is measured in the SI base units of ampere per meter (A/m).[7] B, magnetic flux density, is measured in tesla (in SI base units: kilogram per second2 per ampere),[8] which is equivalent to newton per meter per ampere. H and B differ in how they account for magnetization. In a vacuum, the two fields are related through the vacuum permeability as in Equation 6 but in a magnetized material, the terms differ by the material's magnetization at each point.

$$B = \mu 0H \quad (6)$$

In this paper, the relation between magnetic field and the size of the loops has been explained on the previous section. As the loops being simulated on CST, we know that theoretically the magnetic field is proportionally to the surface area. It means that when the size increases, the magnetic field also increases. From the further simulation, the simulated result obeys the theory. The results will be shown from Figure 5 until Figure 7.

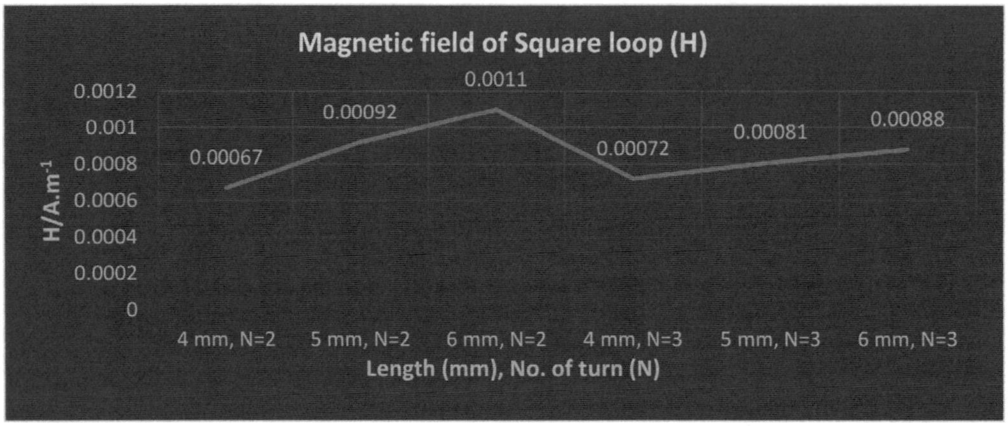

Figure 5. Magnetic field of square loops.

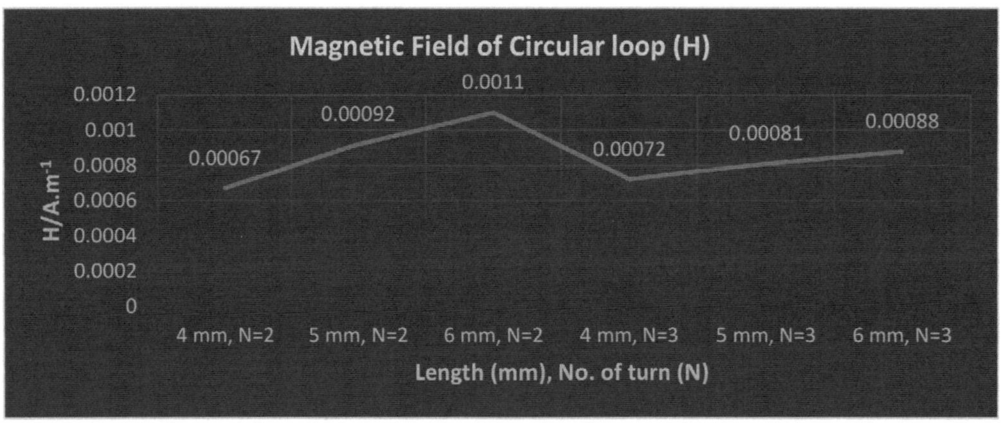

Figure 6. Magnetic field of circular loops.

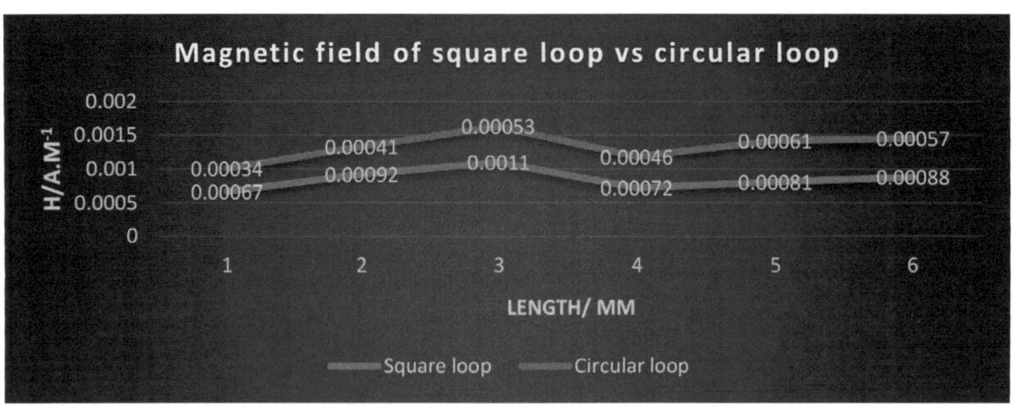

Figure 7. Magnetic field of square and circular loops.

3. Analysis

The inductance of the loop at operating frequency 920 Mhz is 0.22 nH as calculated in [Eq. 3]. Based on the equation, theoretically the inductance value is physically depending on the fill ratio (p) and average diameter of the loop (Davg). The loop width, w and the loop spacing is varied to achieve the inductance value at operating frequency 920 Mhz. The longer the length loop, the longer the length of w and s of the loop and the higher the value of XL. The value of XL must be matched with the value of XC of Monza R6 which is XC= j120.135Ω. Two different vectors are in use to represent a magnetic field: one called magnetic flux density (B) and the magnetic field strength (H). The magnetic field H might be thought of as the magnetic field produced by the flow of current in wires and the magnetic field B as the total magnetic field including also the contribution made by the magnetic properties of the materials in the field.

From the square loop and circular loop simulation, it shows that the higher the value of w and s of the loop, the higher the magnetic field strength (H). However, as the number of the turns increases, the H increases. It can be determined by the magnetic value of two loops of square and circular loop, it has higher magnetic field than the three loops.

4. Conclusion

A single loop IC module is evaluated to fed a wire embedded meandered dipole antenna for UHF RFID tag. A proper matching should be considered to ensure optimum couple energy to turn on the chip. The size reduction does not affect the magnetic field so much and to increase the magnetic flux, number of turns can be added.

5. Reference

[1] D. M. Dobkin, *"The RF in RFID: UHF RFID in Practice,"* 2nd Ed., Elsevier, 2013.

[2] *"Malaysian Communications and Multimedia Commission Act 1998 (Act 589)"* (PDF). Attorney General's Chambers of Malaysia. 1 January 2006. Retrieved 25 April 2018.

[3] H.-W. Son and C.-S. Pyo, *"Design of RFID tag antennas using an inductively coupled feed,"* Electronics Letters, Vols **41**, No.18 , 2005.

[4] Zhao, Jonsenser. (2010). *A new calculation for designing multilayer planar spiral inductors*. Edn-Boston then Denver then Highlands Ranch Co-. **55**. 37-40.

[5] Monza Reference Design Documents & Downloads https://support.impinj.com/hc/enus/sections/200454558-Monza-Reference-Design-Documents-Downloads) – Note: These documents are restricted and require access permission from Impinj.

[6] Feynman, Richard P.; Leighton, Robert B.; Sands, Matthew (1963). *The Feynman Lectures on Physics*. **2**. California Institute of Technology. ISBN 9780465040858.

[7] International Bureau of Weights and Measures (20 May 2019), SI Brochure: *The International System of Units (SI) (PDF)*(9th ed.), ISBN 978-92-822-2272-0, p. 22

Acknowledgments

The authors would like to acknowledge the support from HID GLOBAL Sdn. Bhd., Advance RF and Microwave Research Group, School of Electrical Engineering and Universiti Teknologi Malaysia under grant number 21H69.

Design of a Flexible T-Matched Feed UHF RFID Tag Antenna

NF. Miswadi[1*], NH Abd Rahman[1], S. Subahir[1, *], M.A. Aris[1,2], RA Awang[1], E. H. Lim[3]
[1]Antenna Research Centre (ARC), School of Electrical Engineering, College of Engineering, Universiti Teknologi MARA, 40450 Shah Alam, Selangor, Malaysia.
[2]School of Electrical Engineering, College of Engineering, Universiti Teknologi MARA Cawangan Terengganu Kampus Dungun, Malaysia.
[3]Department of Electrical and Electronic Engineering, Lee Kong Chian Faculty of Engineering and Science, Universiti Tunku Abdul Rahman, Bandar Sungai Long, 43000 Kajang, Selangor, Malaysia.

*Corresponding Author: farahmiswadi@gmail.com

Abstract. A flexible radio frequency identification (RFID) tag antenna with a T-match feeder is proposed in this paper. The T-match feed acts as the conjugate network matching of the tag antenna and 0.99 of power transfer coefficient was achieved. By incorporating a meander line and capacitive tip loading technique, the compact size is realized with a wide bandwidth of 44MHz that can cover the RFID spectrum allocated for the Malaysia region, from 919 to 923 MHz. The proposed antenna is printed on the flexible polyethene terephthalate (PET) substrate with a dimension of 90mm × 24mm × 0.143mm. By employing 4W EIRP power, the detection distance of 29.15m in free space. The proposed meandered line and capacitive tip loading tag antenna with T-match feed line is beneficial to be part of the RFID systems because of their compactness, wide bandwidth and high-power transfer coefficient.

1. Introduction

Radiofrequency identification (RFID) has experienced progressive growth for the past two decades and is broadly implement in various sectors such as supply chain management, item label tagging, healthcare and military applications[1]–[4]. RFID is a well-known automated identification (auto-ID) technology that employs radiofrequency waves to track and identify objects automatically. Unlike other auto-ID technologies such as barcode systems and optical character recognition (OCR), RFID systems have gained attention in commercial applications because they can be tracked from several feet away without requiring a direct line of sight of the reader and higher memory capacity[5]. RFID systems are flexible due to the ability to read data without visual access and from moving objects. Furthermore, its ability to transfer data wirelessly between tag and reader using radio waves will reduce the time to key in the data manually. The efficiency and accuracy of the operation will increase[6], [7].

RFID technology can operate in several frequency ranges: low frequency, high frequency, ultra-high frequency, and microwave. The International Telecommunication Union (ITU) has allocated a UHF-RFID frequency band between 860 MHz and 960MHz[8]. The frequency range allocated for the UHF RFID are between 860 – 920 MHz and vary across the region. Each region has a specific frequency range , i.e., 866 – 869 MHz (Europe), 902 – 928 MHz (US) and 917 – 921 MHz (Japan) [9]. The allocated UHF-RFID frequency band for Malaysia is 919MHz – 923 MHz [10]. RFID systems have four main components: RFID tag which include antenna and chip, reader and computer host to process the data. RFID tag plays an crucial role in the RFID system performance that holds the transmitted data to the reader[11]–[13].

Since the frequency allocated for RFID application is below 1GHz, the size of the tag antenna will be bulky as the antenna size is inversely proportional to the frequency. A big tag antenna cannot cater to the industry needs because it is not suitable to be implemented on small objects. Hence, miniaturization of the tag antenna is one of the aspects that many researchers have been trying to improve. Meandering and capacitive tip loading techniques have been applied to reduce the tag antenna size in [14]. However, as the tag is built on a non-flexible substrate, it is not easy to mount the fabricated antenna on uneven, curved surfaces, so it is not suitable to be implemented at the uneven and curvy surface. Hence it will limit the application in the market. Another approach to miniaturize the size is folding techniques and the tag antenna was folded into three layers to reduce the size as reported in [15]. An artificial Magnetic Conductor (AMC) structure has been embedded into the antenna structure to shrink the footprint of the RFID tag antenna[16]. However, the utilization of the AMC structure and folding technique increases the cost and complexity.

This paper focuses on designing a flexible UHF RFID tag antenna with a meandered line and capacitive tip loading technique to achieve an electrically small tag antenna. Meandering was chosen due to its simple structure and ability to reduce the size by tuning the of the meandered line parameters. Meanwhile, the inductive behaviour of the tag antenna can be realized by introducing the capacitive tip loading[3]. The implementation of the T-match feed as the conjugate matching network is to realize perfectly match impedance and maximum power transfer between RFID chip and tag antenna. Polyethene terephthalate (PET) substrate has been used in this study making the proposed tag antenna application-oriented, low profile and cost-effective in the designing and production process. This paper is prepared in the following way. The proposed tag antenna configuration and parametric study for impedance analysis at different parameters have been provided in Section 2. In Section 3 and 4, performances of tag in terms of input impedance, read range and power transfer coefficient and a conclusion is presented, respectively.

2. Antenna Configuration

In this section, a configuration of the proposed UHF RFID tag antenna is presented, the tag antenna is made up of T-match feed, meandered line and capacitive tip loading that symmetry identity at both sides of the antenna **Figure 1**. The tag antenna is design on PET substrate with a thickness of 12µm, the dielectric constant of 3.4 and tangent loss of 0.003. The RFID chip is integrated and place in the center of the T-match feed structure. NXP Ucode 8 with an impedance of 13 – j 191 Ω and reading sensitivity of - 22.9 dBm have been selected in this paper. The tag antenna was designed and simulated using a CST simulator. The proposed tag antenna can be divided into three regions; Region A, Region B and Region C for conjugate matching network meandered line and capacitive tip loading, respectively.

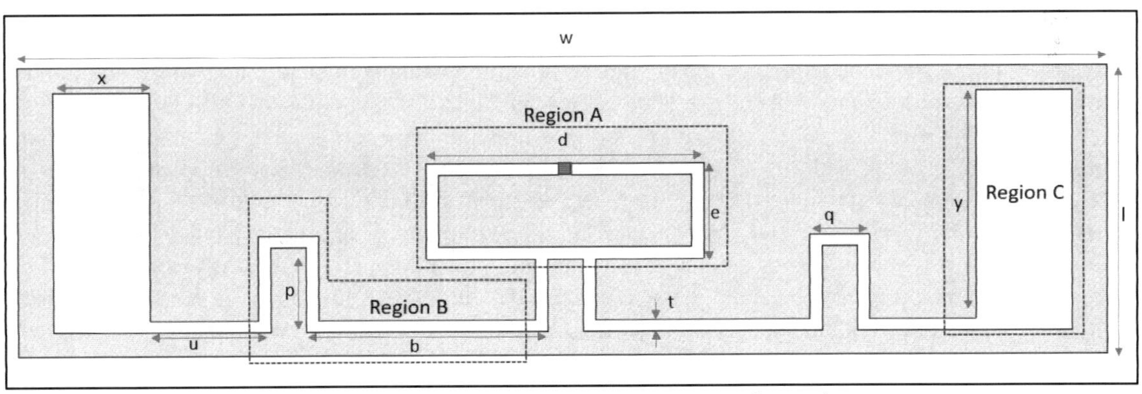

Figure 1. Proposed RFID tag antenna configuration

The conjugate matching network was built to find the ideal match and maximum power transfer of the tag antenna and chip impedance. The impedance of the proposed tag antenna should be 13+j191Ω or nearly that value. The equivalent circuit tag antenna with T-match feed line is depicted in Figure 2 [17]. The input impedance of the proposed antenna in Figure 1, is given as [18]

$$Z_a = R_a + jX_a = Z_t + Z_{ctp} + \frac{(2\pi f M)^2}{Z_m} \tag{1}$$

where Z_t is the T-match feeder impedance, Z_{ctp} is the impedance of the capacitive tip loading, Z_m is the meandered line impedance and M is the inductive mutual coupling between the T-match feeder and radiating body including meandered line and capacitive tip loading.

The transfer power coefficient, τ is given by [19] :

$$\tau = \frac{4R_cR_a}{|Z_c + Z_a|^2} \quad (2)$$

where $Z_c = R_c + jX_c$ is the chip impedance and $Z_a = R_a + jX_a$ is the antenna impedance. The value of the power transfer coefficient should be between 0 to 1.

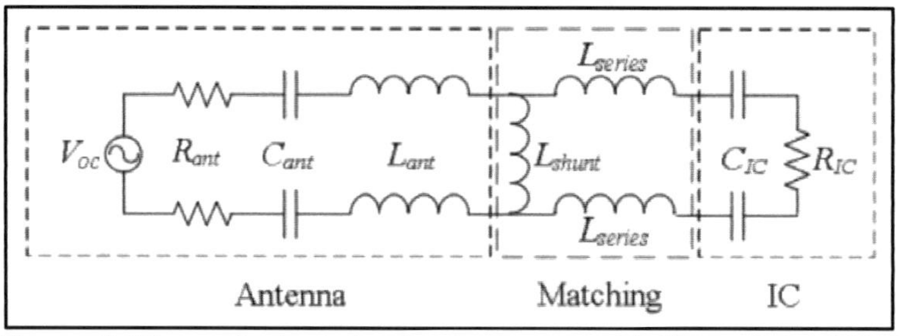

Figure 2. The equivalent circuit tag antenna with T-match feed [17]

3. Result and Discussion

The optimized simulated value of the proposed RFID tag antenna is tabulated in table 1. Figure 3 display the correlation between impedance properties and the dimension of region A of the tag antenna in terms of its wavelength, λ. Region A is the T-match conjugate network structure. By altering parameters *d* and *e,* which are the width and length of the T-match feeder, the inductance values of series and shunt in the equivalent circuit are varied. Increasing the width and length of the T-match feeder causes the value of resistance and reactance to increase and the proposed tag antenna becomes more inductive. The optimum value of the *d* and *e* is found to be 70λ (23mm) and 18λ(6mm). Next, correlation of the between impedance properties and the dimension of meander line while maintaining the width and length of the T-match feeder is shown in **Figure 4**. When the width of the meandered line, *b* is varied from 36λ to 76λ, the of the antenna impedance in terms of resistance and reactance increase from 1Ω to 3Ω and 195 to 197, respectively. The correlation between impedance properties and the dimension of capacitive tip loading of tag antenna is illustrated in **Figure 5**. By tuning the parameter *x* and *y* from 15λ to 33λ and 47λ to 73λ, the result shows that the resistance of the antenna increases, meanwhile the value of the reactance fluctuate. From this parametric study, the resistance or real part of antenna impedance can be easily controlled by tuning the dimension of the radiating body, consisting of meandered line and capacitive tip loading shown by Region B and C, respectively. Meanwhile, by tuning the dimension of the conjugate matching network, shown by the T-match feeder (Region A), the reactance or imaginary part of the antenna impedance are strongly influenced.

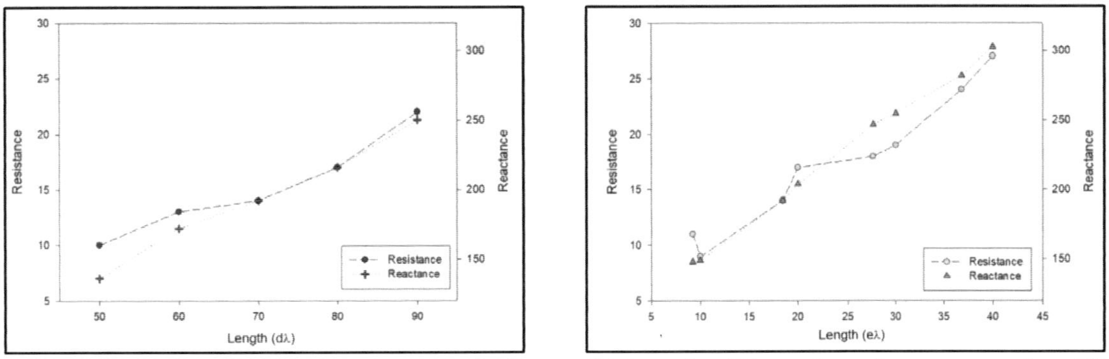

(a) Width of T-match conjugate network, d (λ) (b) Length of T-match conjugate network, e (λ)

Figure 3. Correlation between impedance properties and the dimension of region A of the tag antenna

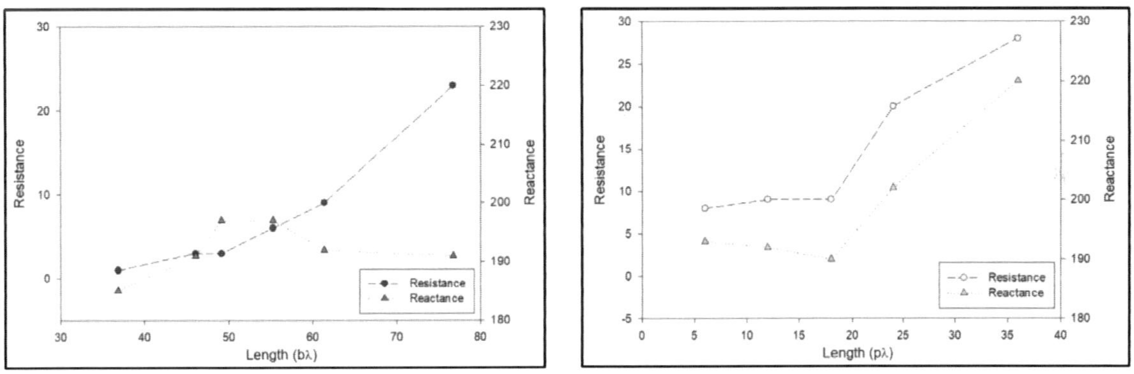

(a) Width of meandered line, b (λ) (b) Length of meandered line, p (λ)

Figure 4. Correlation between impedance properties and the dimension of region B of the tag antenna

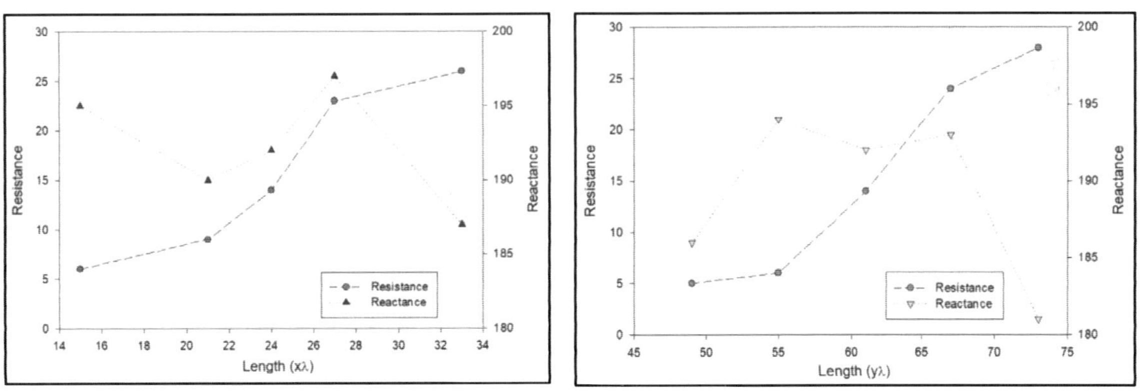

(a) Width of capacitive tip loading, x (λ) (b) Length of capacitive tip loading y (λ)

Figure 5. Correlation between impedance properties and the dimension of region C of the tag antenna

The tag antenna is resonated from 896 MHz to 945 MHz with the centre frequency of 915 MHz as shown in figure 6. The S11 value at 915 MHz is -35.33 dB. The -10 dB bandwidth of the proposed tag antenna is 48 MHz. Figure 7 illustrates the antenna power transfer coefficient and impedance is 0.99 and 14 + j192Ω, which are nearly conjugate to the chip impedance 13 + j191Ω, respectively. As the the power transfer coefficient is 0.99 it indicates that antenna impedance are conjugate with the chip impedance.

Table 1. Parameters and dimensions of the proposed tag antenna

Parameters	Description	Dimension (mm)	Parameters	Description	Dimension (mm)
b	Width of meandered line	20	*x*	Width of capacitive tip loading	8
d	Width of T-match feed	23	*y*	Length of capacitive tip loading	20
e	Length of T-match feed	6	*l*	Length of the tag antenna	24
q	Width of meandered line	5	*w*	Width of the tag antenna	90
p	Length of meandered line	6	*u*	Width of meandered line	10

Read range is the most crucial parameter in the UHF RFID systems because it determines whether the system is efficient or inefficient. Read range is the maximum distance at which the reader can read the data from the tag antenna. By using the Friis free–space formula, the maximum read range of the tag antenna can be calculated using equation [20]:

$$R = \frac{\lambda}{4\pi}\sqrt{\frac{P_t G_t G_{tag}\tau}{P_r}} = \frac{\lambda}{4\pi}\sqrt{\frac{P_t G_t G_r}{P_{c,on}}} \quad (3)$$

where $P_{eirp} = P_t G_t$ is the maximum transmitting power allowed (4W) and $P_{c,\,on}$ is the threshold power to turn on the chip or chip sensitivity. Based on the chip and system specification, the calculated read range of the RFID system is 29.15m, given that the simulated realized gain of the proposed tag antenna is 2.059 dBi. The surface currents distribution of the proposed tag antenna is shown at Figure 8. The high current density can be seen at the centre of the tag antenna structure as the T-match feeding structure located.

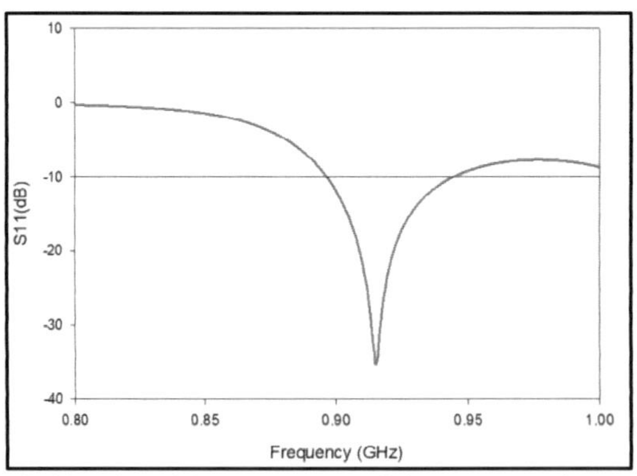

Figure 6. Simulated S-parameter of the proposed tag antenna

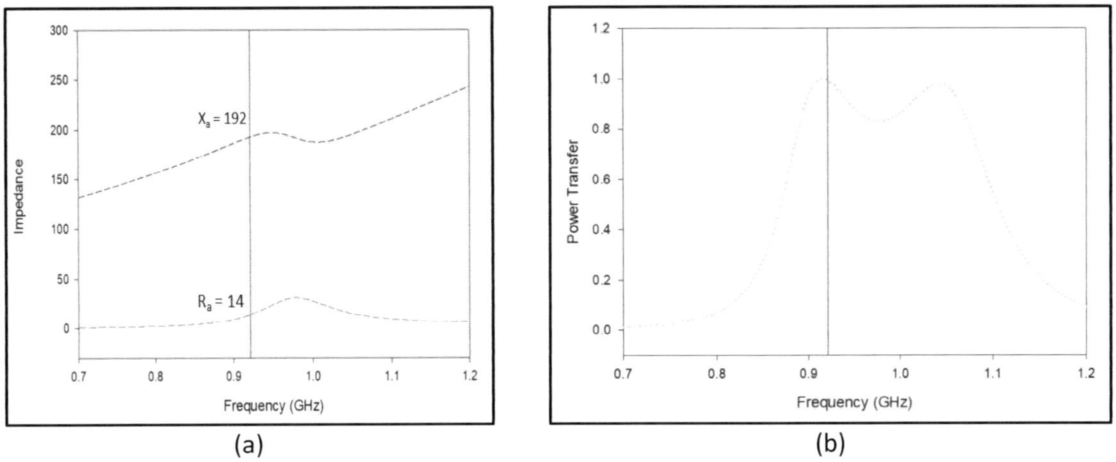

Figure 7. (a) Impedance (b) Power Transfer Coefficient of the proposed tag antenna

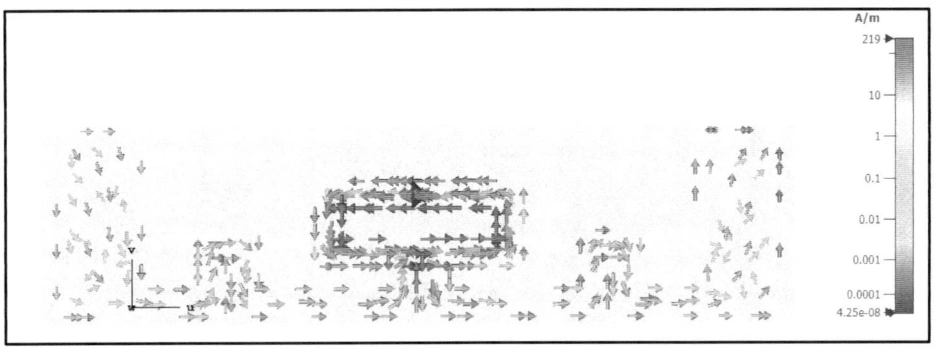

Figure 8. Surface current distribution of the proposed tag antenna

4. Conclusion

A flexible UHF RFID tag antenna with a total size of 90mm × 24mm × 0.125mm is presented in this paper. By employing the meandering technique, capacitive tip loading and T-match conjugate network, the size of the antenna has been reduced and the antenna impedance is matched and achieved the maximum power transfer of 0.99. The proposed antenna has a wide bandwidth covering the whole RFID frequency band for the Malaysia region.

References

[1] N. A. S. Abdullah, N. H. Ramli, and R. C. Yob, "A RFID Tag Antenna for Wearable Application System in Microwave Frequency Band," *J. Phys. Conf. Ser.*, vol. 1962, no. 1, 2021, doi: 10.1088/1742-6596/1962/1/012034.

[2] S. Wang, Y. Liu, and T. T. Ye, "Unconventionally Shaped" Antenna Design for UHF RFID Tags," *Int. J. Antennas Propag.*, vol. 2021, 2021, doi: 10.1155/2021/9965252.

[3] M. Hussain, Y. Amin, and K. G. Lee, "A compact and flexible UHF RFID tag antenna for massive IoT devices in 5G system," *Sensors (Switzerland)*, vol. 20, no. 19, pp. 1–21, 2020, doi: 10.3390/s20195713.

[4] M. Kumar, A. Sharma, and I. J. G. Zuazola, "All-in-One Uhf Rfid Tag Antenna for Retail Garments Using Nonuniform Meandered Lines," *Prog. Electromagn. Res. Lett.*, vol. 94, no. November, pp. 133–139, 2020, doi: 10.2528/pierl20072102.

[5] J. Vijayalakshmi, V. Dinesh, P. A. Sivsankgari, S. Saranya, and V. Sanju, "A Ultra High Frequency (UHF) RFID Antenna Design for Food Quality and Safety," *Int. J. Recent Technol. Eng.*, vol. 8, no. 6, pp. 1013–1022, 2020, doi: 10.35940/ijrte.f7566.038620.

[6] S. Surwase, R. Yadahalli, and S. Nawale, "Design of Nested H slot Passive UHF RFID Tag," *Int. J. Eng. Adv. Technol.*, vol. 10, no. 1, pp. 346–349, 2020, doi: 10.35940/ijeat.c4721.1010120.

[7] R. Abdulghafor *et al.*, "Recent Advances in Passive UHF-RFID Tag Antenna Design for Improved Read Range in Product Packaging Applications: A Comprehensive Review," *IEEE Access*, vol. 9, pp. 63611–63635, 2021, doi: 10.1109/ACCESS.2021.3074339.

[8] Radiocommunication Sector of ITU, "Technical characteristics, standards, and frequency bands of operation for radio-frequency identification (RFID) and potential harmonization opportunities," 2012. [Online]. Available: https://www.itu.int/dms_pub/itu-r/opb/rep/R-REP-SM.2255-2012-PDF-E.pdf.

[9] M. T. Islam, T. Alam, I. Yahya, and M. Cho, "Flexible radio-frequency identification (RFID) tag antenna for sensor applications," *Sensors (Switzerland)*, vol. 18, no. 12, 2018, doi: 10.3390/s18124212.

[10] MCMC, "SPECIFICATION FOR SHORT RANGE DEVICES (SRD)," 2014.

[11] Y. Gmih and A. Farchi, "Compact Antenna for UHF-RFID Tag Tested on the Human Body for Identification Cards," *Int. J. Intell. Eng. Syst.*, vol. 13, no. 1, pp. 1–10, 2020, doi: 10.22266/IJIES2020.0229.21.

[12] Y. El Hachimi and A. Farchi, "Analysis, design, and manufacture of a compact UHF RFID tag antenna for operating in a metallic or non-metallic environments," *Int. J. Intell. Eng. Syst.*, vol. 13, no. 1, pp. 282–290, 2020, doi: 10.22266/ijies2020.0229.26.

[13] K. Mekki, "Design of a miniaturized dipole RFID tag antenna," *Indian J. Sci. Technol.*, vol. 13, no. 30, pp. 2203–2212, 2020, doi: 10.17485/ijst/v13i30.114.

[14] A. A. S. A. Shah, N. H. A. Rahman, M. T. Ali, and M. R. Ahmad, "Design of Electro-Textile UHF RFID Tag through T-Match Network," *Proc. IEEE Int. RF Microw. Conf.*, pp. 57–60, 2018, doi: 10.1109/RFM.2018.8846547.

[15] S. Y. Ooi, E. H. Lim, P. S. Chee, Y. H. Lee, K. Y. Lee, and F. L. Bong, "Inverted Patch with an Inductive Loop-Shaped Feeder for On-Metal Tag Design," *2020 IEEE Int. RF Microw. Conf. RFM 2020 - Proceeding*, pp. 23–25, 2020, doi: 10.1109/RFM50841.2020.9344738.

[16] N. Ripin, E. H. Lim, F. L. Bong, and B. K. Chung, "Miniature Folded Dipolar Patch with Embedded AMC for Metal Mountable Tag Design," *IEEE Trans. Antennas Propag.*, vol. 68, no. 5, pp. 3525–3533, 2020, doi: 10.1109/TAP.2020.2969814.

[17] N. M. Faudzi, A. N. A. Rashid, A. Ibrahim, M. F. Khyasudeen, and M. T. Ali, "Microstrip Dipole UHF-RFID Tag Antenna for Metal Object Tagging," *Proc. - 6th Int. Conf. Comput. Commun. Eng. Innov. Technol. to Serve Humanit. ICCCE 2016*, no. July, pp. 36–41, 2016, doi: 10.1109/ICCCE.2016.21.

[18] F. Erman *et al.*, "U-Shaped Inductively Coupled Feed UHF RFID Tag Antenna with DMS for Metal Objects," *IEEE Antennas Wirel. Propag. Lett.*, vol. 19, no. 6, pp. 907–911, 2020, doi:

10.1109/LAWP.2020.2981960.

[19] X. Xuan, L. Lv, and K. Li, "A Miniaturized Meandered Dipole UHF RFID Tag Antenna for Flexible Application," *Int. J. Antennas Propag.*, vol. 2016, 2016, doi: 10.1155/2016/2951659.

[20] F. K. Byondi and Y. Chung, "Longest-range UHF RFID sensor tag antenna for iot applied for metal and non-metal objects," *Sensors (Switzerland)*, vol. 19, no. 24, 2019, doi: 10.3390/s19245460.

Acknowledgments
This author would like to express gratitude to the researchers of Antenna Research Center (ARC), College of Engineering, Universiti Teknologi MARA for their guidance. This research has been supported by UiTM under a research grant with the grant number of 600-RMC/MYRA 5/3/LESTARI (066/2020).

Journal of Physics: Conference Series 2250 (2022) 012016 doi:10.1088/1742-6596/2250/1/012016

Utilizing Logistic Map to Enhance the Population Diversity of PSO

S M Ajibade[1], **R Chweya**[2], **M O Ogunbolu**[3], **S R Fadipe**[4]

[1] Dept. of Computer Engineering, Istanbul Ticaret Universitesi, Istanbul, Turkey
[2] School of Information Science and Technology, Kisii University, Kisii, Kenya
[3] Dept. of Computer Science, Lead City University, Ibadan, Oyo State, Nigeria
[4] Lagos State University, Ojo, Lagos, Nigeria. Orcid id: 0000-0003-0695-5193

Corresponding author: asamuel@ticaret.edu.tr

Abstract. Particle swarm optimization (PSO) is a high-quality, nature-inspired global optimization algorithm that can be applied to a variety of real-world optimization problems. PSO, on the other hand, has some flaws, such as slow convergence, premature convergence, and the ability to become stuck at local optimum solutions. This research aims to address the issue of population diversity in the PSO search process, which leads to premature convergence. As a result, in this study, a method is introduced to PSO in order to avoid early stagnation, which leads to premature convergence. A chaotic dynamic weight particle swarm optimization (CTPSOA) is proposed, in which a chaotic logistic map is delivered to increase the population range within the PSO search technique by editing the inertia weight of PSO to avoid premature convergence. This study also looks into the overall performance and viability of the proposed CTPSOA as a set of function selection rules for solving optimization issues. There are eight traditional benchmark functions that are used to assess the overall results and obtain the accuracy of the proposed (CTPSOA) algorithms when compared to a few other meta-heuristics optimization rules. The test results reveal that the CTPSOA outperforms other meta-heuristics algorithms in solving optimization problems by 85% and has established itself as a reliable and superior metaheuristics algorithm for feature selection.

1. Introduction

"Data mining is a technique of removing and finding styles in enormous datasets concerning strategies on the crossing point of AI, realities, data set designs. Data Mining techniques areused to find stowed away plans and relations on a concentrated proportion of realities that is most likelygainful in settling on of decisions" [1]. "Because the majority of the datasets created contain large amounts of redundant and non - significant elements, there may be a significant decrease in the precision with which the classifiers are prepared, as well as a reduce in the classifier's training frequency of the classifiers" [2]. "This problem is known as the curse of dimensionality in information mining methods, and it increases the computational complexity of developing classifiers" [3].

"The total space of exploring is made up of all possible subsets of capabilities, indicating that the degree of the request space is 2n, where n is the number of actual capabilities" [4]. "Metaheuristics-based techniques have recently gotten a lot of attention because of their great result in tackling the feature selection challenges and also because of their future global search ability" [2]. "Slow convergence, pre-mature convergence, and easily becoming stuck at neighborhood most efficient approaches are some of PSO's flaws" [3]. To address these flaws, (CTPSOA) is proposed, in which a logistic map is added to increase the population range within the PSO seek procedure by increasing the inertia weight of PSO.

2. Related Studies

2.1. Particle Swarm Optimization (PSO)

"PSO is an optimization algorithm created by Kennedy and Eberhart in 1995" [6]. "It mimics the social behavior of individuals in swarms of fish and fowls. PSO has been shown to be less computationally expensive and to easily integrate with other meta-heuristics techniques such as Genetic Algorithm (GA), Ant Colony Optimization (ACO), Differential Evolution (DE)" [2], among others. "PSO is really not complex to implement, requiring only a few fine-tuning parameters, and it requires significantly less memory and processing power. Consequently, it is commonly used in a number of domains as such an efficient strategy for the selection of feature" [2]. Regardless of the fact that the PSO is incredibly adept at searching for ideal components subsets, it suffers from a variety of issues, such as premature convergence, which is common in complex optimization problems [5], and they also overlook the volume of components in their search while focusing solely on reducing classification error rates. As shown in equations (1) and (2), "By resetting the flying velocity and neighborhood of each person in the swarm, the PSO method differentiates the global ideal results" [2]:

$$v_{id}^{(t+1)} = v_{id}(t) + b_1(t) \times d_1 \times (pbest_{id}^{(t)} - x_{id}^{(t)}) + b_2(t) \times d_2 \times (gbest^{(t)} - x_{id}^{(t)}) \quad (1)$$

$$x_{id}^{(t+1)} = x_{id}^{(t)} + v_{id}^{(t+1)} \quad (2)$$

where b_1 and b_2 are two uniformly distributed values in the range [0,1], and d_1 and d_2 are acceleration parameters, which are commonly set to 2.0. As a result of the use of the inertia weight, equation (1) has been improved to equation (3), as shown below:

$$V_{id}^{(t+1)} = \omega + v_{id}(t) + b_1(t) \times d_1 \times (pbest_{id}^{(t)} - x_{id}^{(t)}) + b_2(t) \times d_2 \times (gbest^{(t)} - x_{id}^{(t)}) \quad (3)$$

2.2. Inertia Weight

"The inertia weight of is controlled using the chaotic map. Tweaking the inertia weight of PSOs utilizing chaotic maps could really raise the likelihood of PSO escape from local neighbourhood optimum solution while resolving multimodal functions, so the logistic map is selected to tweak inertia weight of PSO" [7]. "The efficient decisions, anomaly, and non-repetition characteristics of the logistic map have the potential to increase wide range of population, and also enhance the functioning of converging at global optimum, avoiding untimely convergence" [8]. It is shown in equation (4).

$$\omega = \varphi * cos\ (\frac{Mj}{Mmax}) * \pi) + \boldsymbol{\tau} \quad (4)$$

2.3. Chaotic Logistic Map

"Innovative, quasi, and limited systems all exhibit chaos, which is a predetermined, irregular technique. Chaos is a statistically occurring simple but spontaneous causal computational modeling structure, as well as the system that involves chaos is regarded as a source of randomness" [5]. "Chaotic maps influence the convergent rate of algorithms positively because they incorporate chaos in the realistic abreast and seem to be for only a brief duration and randomness for a predefined timeframe" [6].

3. Methodology

All of the processes that are involved in achieving the research objectives are contained in the research framework. There are two stages to this study's framework. The framework's first phase described the method for performing pre-processing of data while the second phase 2 described the PSO Initialization which included the introduction of a chaotic logistic map as seen in Figure 1.

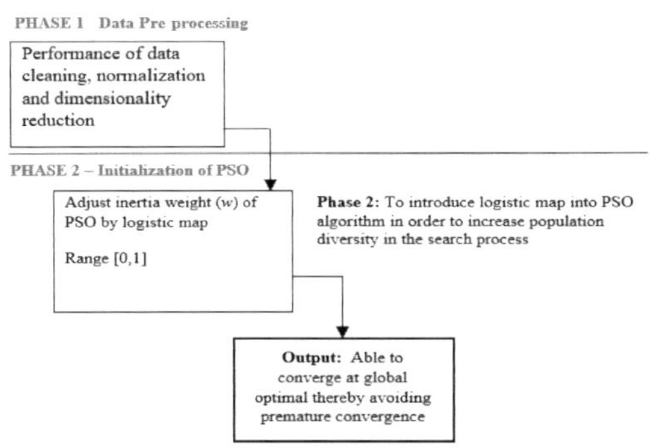

Figure 1. Research Operational Framework

Three fundamental adjustments are proposed to improve PSO performance, including merging the dynamic weights, acceleration coefficient, and best-so-far position to refresh the new situation with the other two sections. Equation includes two distinct weight adjustment factors, denoted by the letters s_1 and s_2 (2). The location of every swarm particle is refreshed by means of the proposed equation (5).

$$x_{id}(t+1) = s_1(t) \text{ x } x_{id}(t) + s_2(t) \text{ x } v_{id}(t+1) + \rho \text{ x } gbest^d \text{ x } \psi \quad (5)$$

As displayed in equation (5), these two boundaries (s_1 and s_2) can handle the compromise between exploitation and exploration, this way they regulate the impact of the previous resolution $x_{id}(t)$ and the speed $v_{id}(t+1)$, $gbest^d$ is the best so far position which speeds up the rate of convergence, and ψ is the acceleration coefficient that manages the greatest advance size and ρ is an irregular number somewhere in the range of 0 and 1. Elements s_1, s_2 and ψ are described as:

$$s_1(t) = \psi = \frac{1}{(1+\exp(a.(-\min(SP)/\max(SP))))t} \quad (6)$$

$$s_2(t) = 1 - s_1(t) \quad (7)$$

The proposed algorithm is shown in figure 2, step 9 shows the enhancement made to the algorithm in that the position of the particle velocity was updated by adjusting the inertia weight using equation 7 and step 12 also highlights the enhancement made to the algorithm in that the particle position of each particle was updated by the proposed update equation using equation 6.

```
1:  Input: NP, D, c1, c2, ω, Mmax, Vmin, Vmax
2:  Randomly initialize each particle position Xi (i=1,2, ..., NP) in the swarm
3:  Generate the initial velocities Vi (i=1,2, ..., NP) for each particle randomly
4:  Calculate the fitness value of each particle using the objective function f
5:  Set pbest and gbest in the swarm
6:  Update the particle velocity Vi using Equation 3
7:  Update the particle position Xi using Equation 2
8:  Chaotic search K times near gbest
9:  Update the position of particle velocity by adjusting inertia weight with chaotic
    logistic map
10: K chaotic search points near pbesti are obtained
11: Calculate weight factors s1 and s2 using equation (6) and (7)
12: Update particle position of each particle by
    proposed update equation
13: end
```

Figure 2. The Proposed CTPSOA Algorithm

4. Experimental Results

"The Number of Function Evaluation (NFE) is utilized as a basis to investigate the convergence rate pf algorithms" [9]. The Pop size (NP) is 50, while the dimension set to 30 and 50, according to [10]. When the value of NFE is small, then it depicts a quicker rate of convergence. Figure 3 also depicts the numerical benchmark functions (f1–f8). Table 1 shows the avg NFE of CTPSOA algorithms for 8 test functions in dimension 30.

Table 1. Avg NFE of CTPSOA algorithm in D=30

Function	GA	PSO	CTPSOA
*f*1	20014	16984	**7124**
*f*2	20019	17169	**3879**
*f*3	20013	20001	**9898**
*f*4	14574	17146	**8145**
*f*5	12491	18633	**6635**
*f*6	20000	20001	**16252**
*f*7	20055	8226	**10163**
*f*8	13682	20001	**6689**

No	Function	Formula	Value	Dim	Range	Properties
1	Sphere	$f(x) = \sum_{i=1}^{n} x_i^2$	0	30	[-100, 100]	Unimodal, Separable
2	Rosenbrock	$f(x) = \sum_{i=1}^{n-1} [100(x_{i+1} - x_i^2)^2 + (x_i - 1)^2]$	0	30	[-30, 30]	Unimodal, Inseparable
3	Quartic	$f(x) = \sum_{i=1}^{n} ix_i^4 + random[0,1)$	0	30	[-1.28, 1.28]	Unimodal, Inseparable
4	Griewank	$f(x) = \frac{1}{4000} \sum_{i=1}^{n} x_i^2 - \prod_{i=1}^{n} \cos \frac{x_i}{\sqrt{i}} + 1$	0	30	[-600, 600]	Multimodal, Inseparable
5	Sumsquare	$f(x) = \sum_{i=1}^{n} ix_i^2$	0	30	[-5.12, 5.12]	Unimodal, Separable
6	Schwefel	$f(x) = \sum_{i=1}^{n} -x_i \sin\left(\sqrt{\lvert x_i \rvert}\right)$	0	30	[-500, 500]	Multimodal, Separable
7	Rastrigin	$f(x) = \sum_{i=1}^{n} x_i^2 - 10\cos(2\pi x_i) + 10$	0	30	[-5.12, 5.12]	Multimodal, Separable
8	Michalewicz	$f(x) = -\sum_{i=1}^{n} \sin(x_i)(\sin(ix_i^2/\pi))^{2m}$	-4.688	5	[0, π]	Multimodal, Separable

Figure 3. Numerical Benchmark Functions

Since the PSO population diversity has been was increased for CTPSOA to obtain a higher effectiveness. The proposed CTPSOA algorithm has a vastly greater rate of convergence than other algorithms as displayed in Table 1.
Table 1 shows the avg NFE of CTPSOA algorithms for 8 test functions in dimension 50.

Table 2. Avg NFE of CTPSOA algorithm in D=50

Function	GA	PSO	CTPSOA
*f*1	20025	11702	**10905**
*f*2	20016	12525	**8191**
*f*3	20013	20000	**19948**
*f*4	18308	18627	**18462**
*f*5	19536	20000	**8239**
*f*6	18999	20000	**20001**
*f*7	20018	**1335**	13711
*f*8	16941	20000	**19976**

Since the PSO population diversity has been was increased for CTPSOA to obtain a higher effectiveness. The proposed CTPSOA algorithm has a vastly greater rate of convergence than other algorithms as displayed in Table 2.

Table 3. Avg Mean Fitness and S.D. of GA, PSO and CTPSOA in D = 30

Function	GA		PSO		CTPSOA	
	Fitness	S.D	Fitness	S.D	Fitness	S.D
*f*1	3.07E-20	0.49E-10	0.4E-31	1.76E-17	**6.74E-14**	**0.318**
*f*2	0.7552	2.18E-04	**6.52E-05**	**0.276030**	5.8E-12	2.63E-11
*f*3	0.42E-13	6.8E-13	0.1058	112.7756	**228.6881**	**0.307501**
*f*4	2.380552	4.081682	0.287015	3.054788	**1.144336**	**1.005037**
*f*5	0.420534	0.457355	0.040014	0.700785	**4.502862**	**0.170012**
*f*6	0.057728	0.031688	0.02635	1.510484	**0.043424**	**0.028741**
*f*7	3.37E-01	1.34E-02	1E-05	2.76E-05	**3.86E-05**	**8.14E-05**
*f*8	5.51E-05	1.86E-05	0.218134	0.780746	**4.28E-04**	**7.88E-04**

8 test functions are being utilized, as shown in Table 3. The outcome shows that CTPSOA outperformed the original PSO in terms of convergence to global optimum.

Table 4. Avg Mean Fitness and S.D. of GA, PSO and CTPSOA in D = 50

Function	GA		PSO		CTPSOA	
	Fitness	S.D	Fitness	S.D	Fitness	S.D
*f*1	6.66E-06	1.10E05	2.42E-05	4.0864	**2.64E-07**	0.62E06
*f*2	3.06E-07	2.34E-05	60.07030	12.16580	**0.521E-05**	**1.33E-05**
*f*3	**1.783361**	**2.358087**	1832.530	1004.638	261.7877	701.6083
*f*4	11.2834	14.03024	115.0206	55.1868	**7.268306**	**1.100604**
*f*5	3.66470	2.073868	36.30812	9.81864	**1.216202**	**4.200553**
*f*6	0.33613	0.067208	37.33570	14.84074	**0.03074**	**0.068854**
*f*7	3.7E-07	1.50E-06	3.73E-05	0.000380	**0.00051**	**3.03E-10**
*f*8	0.50344	0.564055	0.000763	0.235181	**3.37E-07**	**3.47E-07**

8 test functions are being utilized, as shown in Table 4. The outcome shows that CTPSOA outperformed the original PSO in terms of convergence to global optimum.

5. Conclusion

Finally, as seen in the findings, the proposed CTPSOA method seems to be more accurate and convenient than GA and PSO at preventing premature convergence. Since it possesses swarm diversity that is poor, the PSO exhibit its weaknesses when trying to resolve the unimodal and multimodal issues. The logistic map is incorporated into PSO to modify the inertia weight, hence enhancing the speed of convergence and preventing premature convergence.

Reference

[1] Ajibade, S. S. M., Ahmad, N. B. B., & Shamsuddin, S. M. (2019). A Novel Hybrid Approach Of Adaboostm2 Algorithm And Differential Evolution For Prediction Of Student Performance. *International Journal of Scientific and Technology Research, 8 (07), 65, 70.*

[2] Makinde, O., & Chakraborty, B. (2018). On some classifiers based on multivariate ranks. *Communications in Statistics-Theory and Methods*, *47*(16), 3955-3969.

[3] Azmi, M. S., Arbain, N. A., Muda, A. K., Abas, Z. A., & Muslim, Z. (2015, November). Data normalization for triangle features by adapting triangle nature for better classification. In *2015 IEEE Jordan Conference on Applied Electrical Engineering and Computing Technologies (AEECT)* (pp. 1-6). IEEE.

[4] Ajibade, S. S. M., Ahmad, N. B., & Shamsuddin, S. M. (2019, August). An heuristic feature selection algorithm to evaluate academic performance of students. In *2019 IEEE 10th Control and System Graduate Research Colloquium (ICSGRC)* (pp. 110-114). IEEE.

[5] Ajibade, S. S. M., Ahmad, N. B. B., & Zainal, A. (2020, July). A Hybrid Chaotic Particle Swarm Optimization with Differential Evolution for feature selection. In *2020 IEEE Symposium on Industrial Electronics & Applications (ISIEA)* (pp. 1-6). IEEE.

[6] Kennedy, J., & Eberhart, R. (1995, November). Particle swarm optimization. In *Proceedings of ICNN'95-international conference on neural networks* (Vol. 4, pp. 1942-1948). IEEE.

[7] Niu, P., Chen, K., Ma, Y., Li, X., Liu, A., & Li, G. (2017). Model turbine heat rate by fast learning network with tuning based on ameliorated krill herd algorithm. *Knowledge-Based Systems*, *118*, 80-92.

[8] Wang, G.-G., Guo, L., Gandomi, A. H., Hao, G.-S., & Wang, H. (2014). Chaotic krill herd algorithm. Information Sciences, 274, 17-34.

[9] Campos, J. A., Jiménez-Bello, M., & Alzamora, F. M. (2020). Real-time energy optimization of irrigation scheduling by parallel multi-objective genetic algorithms. *Agricultural Water Management, 227*, 105857.

[10] Verma, O. P., Manik, G., & Sethi, S. K. (2019). A comprehensive review of renewable energy source on energy optimization of black liquor in MSE using steady and dynamic state modeling, simulation and control. *Renewable and Sustainable Energy Reviews*, *100*, 90-109

AUTHOR INDEX

Institute of Physics
Dirac House, Temple Back
Bristol BS1 6BE UK

ISSN: 1742-6588
ISBN 978-1-7138-5504-0